$29.95

Finding Birds

in Darwin, Kakadu and the Top End

Northern Territory
Australia

D1767013

Niven McCrie   James Watson

FINDING BIRDS
IN DARWIN, KAKADU
AND THE TOP END

by Niven McCrie and James Watson

© 2003 Niven McCrie and James Watson

All rights reserved. No part of this publication may
be reproduced or transmitted in any form or by any
means without the prior written consent of the
publisher.

Publisher: Niven McCrie
Design & Layout: Niven McCrie
Maps & Illustrations: Niven McCrie

NATIONAL LIBRARY OF AUSTRALIA, CATALOGUING-IN-PUBLICATION ENTRY
McCrie, Niven.
Finding birds in Darwin, Kakadu and the Top End,
Northern Territory, Australia.
Bibliography.
ISBN 0 646 42183 2.
1. Birds - Northern Territory - Identification.
2. Bird watching - Northern Territory - Guidebooks.
I. Watson, James, 1977- . II. Title.
598.0994295

First Published in 2003 by NT Birding
P.O. Box 41382
Casuarina
Northern Territory
Australia

Contents

Acknowledgements

Some of the sites described in this book are traditional birdwatching areas and there is no knowing who first discovered them. When he came to Darwin in January 1985, Niven was introduced to many of them by Tony Hertog, John McKean, Keith and Lindsay Fisher, Hilary Thompson, Johnny Estbergs and Fred van Gessel. He expresses his thanks to those people for the generous way in which they shared their knowledge of the Top End, its birds and its birdwatching sites.

Many people helped with field testing the various drafts of texts and maps and offered useful comments on them. Others contributed their own observations and notes on the sites in the book. We are grateful to David Donato, Richard Noske, Stuart Robertson, Don Franklin, Nick Leseberg, Bruce Michael, Maria Bellio, Mat Gilfedder, Robbie Robinson, Wendy Hare, Rob Farnes, David Harper, Adrian Boyle, Gavin O'Brien, Mike Ostwald, Arthur and Sheryl Keates, Bryan Baker, Peter Wilkins and staff of the NT Parks and Wildlife Commission for these important contributions.

For their help in the extensive tasks of reviewing the various drafts of this book the authors are especially grateful to David Donato, Richard Noske, Don Franklin and Iris Beale. We also thank Roger Jaensch, Tom McCrie, Stuart Robertson, Judy Evans and Trevor Ford for their helpful comments on the manuscript.

Cathy Robinson kindly undertook the job of senior editor. We appreciate the commitment and skills she brought to the task and know that it is all the more readable as a result.

Stuart Robertson, Johnny Estbergs and Adrian Boyle generously provided photographs for the cover of the book.

James would like to thank his parents, Bill and Sandra, for teaching him an appreciation of birds from a young age and his brothers, Richard and Alexander, for sharing his enthusiasm for Australian wildlife. James would also like to thank his colleagues and teachers at the Australian Defence Force Academy and the University of Oxford for focusing his love of birds into applied research and a passion for conservation.

Niven appreciates the important role played by his son Tom. His interest in birding as he grew up provided, and continues to provide, a reason to keep revisiting the old sites and to find new ones to 'pin down' the more elusive Top End birds. Niven wishes to express his deepest gratitude to his partner Judy for her unflinching support and understanding of all the time and energy he has put into writing this book.

Introduction

The Top End of Australia is one of the continent's birdwatching 'jewels'. Many birdwatchers are drawn to the Top End because of its rich species diversity, with more than 350 species of birds recorded for the region. Five species (Partridge Pigeon, Chestnut-quilled Rock-Pigeon, Banded Fruit-Dove, Hooded Parrot and White-throated Grasswren) are endemic to the region and a further ninety species found in the Top End are endemic to Australia. Moreover, many birds occurring throughout the wet-dry tropics of northern Australia can be seen much more easily in the Top End than elsewhere. Unfortunately, visiting birdwatchers are often disappointed with the lack of reliable information available on where species may be found.

We had two aims for writing this book. The first aim was to help local and visiting birdwatchers find their target bird species, and detailed maps and site descriptions of a large number of accessible birdwatching sites within the Top End are a major feature of this book. Even though some species are very rare in the region, we have ensured that there is at least one site for every resident and migratory bird species within the Top End. The second aim was to contribute to, and build on, the current knowledge of the distribution and status of indigenous bird species within the Top End. By documenting our own extensive observations in the region over many years, supplementing these data with the observations of many other birdwatchers and conducting an extensive review of the literature, we have produced the first accurate annotated list of birds of the Top End (see below for area covered). This list describes each species' abundance, distribution, seasonality of occurrence and preferred habitat in the region. We hope that as a consequence of this work, birdwatchers will be able to more easily find their 'target' bird species and, in turn, further add to what is known about the status and distribution of birds in the region.

The Area & When to Visit

For our purposes, the Top End is that part of the Northern Territory north of 16° S latitude (see map, inside front cover). The area is bounded to the south by Warloch Ponds and by the Victoria Highway, which runs southwest from Katherine. Keep River National Park, at the Northern Territory - Western Australia border, forms the western boundary. Much of the eastern Top End is Aboriginal land and not readily accessible to the casual birdwatcher, so this book only covers the mainland areas of the western Top End. Fortunately all of the Top End species can be found in the accessible areas. While there is no *best time* for birding in the Top End, the presence of particular migratory species (p. 8) and the weather (p. 10) will be important considerations in planning when to visit. The highest number of species are present in September/October, though the weather is more pleasant in June and July. During the wettest months, December - March, access to many areas becomes restricted and tropical downpours frequently curtail daily birdwatching.

Birding Habitats in the Top End

Wetlands

Wetlands comprise swamps, lagoons, lakes, floodplains and artificial waterbodies such as sewage ponds and farm dams. Natural wetlands in the Top End fill with water during the Wet Season, then gradually dry up from May onwards. Waterbirds move toward the coast as wetlands further inland dry up. Pink-eared Duck and Hardhead, for example, build up in large numbers on the coast if the inland is particularly dry. Conversely, during the Wet Season, waterbirds disperse to subcoastal and inland areas.

Shores

The inshore waters, beaches, reefs and coastal mudflats of the region are dominated by shorebirds, terns and herons. True seabirds are very scarce in the Top End, though frigatebirds and boobies may occur from October to April.

Mangroves

Extensive areas of mangroves grow along the coasts and estuaries of the Top End. Several species of birds including Chestnut Rail, Mangrove Grey Fantail and Mangrove Robin are restricted to this habitat. Others such as Yellow White-eye, Red-headed Honeyeater and Black Butcherbird are predominantly found in mangroves but may also be seen in adjacent habitats.

Monsoon forest

The 'rain forests' of the wet-dry tropics, sometimes called Monsoon Vine Forests and Thickets, are mainly restricted to coastal areas, the edges of wetlands and along rivers and streams. Orange-footed Scrubfowl, Rose-crowned Fruit-Dove and Rainbow Pitta are characteristic birds of this habitat.

Open woodland

Eucalypt woodland with a grass understorey is the most extensive habitat in the Top End. Typical woodland birds include parrots, honeyeaters and finches, many of which undertake seasonal movements, following flowering trees and seeding grasses.

Sandstone plateaux and escarpments

Sandstone plateaux and escarpments occur in Kakadu and Gregory National Parks. Within their range, Chestnut-quilled and White-quilled Rock-Pigeons, Banded Fruit-Dove, White-throated Grasswren, Sandstone Shrike-thrush and White-lined Honeyeater are all endemic to this habitat.

Seasonality of Birds in the Top End
Wet Season Migrants

The Wet Season migrants come mainly from the northern hemisphere (Tables 1 & 2). Most are shorebirds, waterbirds or non-passerine landbirds, arriving from August to November and departing from March to May.

Table 1. Palearctic shorebirds recorded in the Top End.

Latham's Snipe [3]	Common Sandpiper	Stilt Sandpiper [3]
Pin-tailed Snipe [3]	Grey-tailed Tattler [4]	Broad-billed Sandpiper [2]
Swinhoe's Snipe	Ruddy Turnstone	Ruff [2]
Black-tailed Godwit	Asian Dowitcher [2]	Red-necked Phalarope [2]
Bar-tailed Godwit [4]	Great Knot	Pacific Golden Plover
Little Curlew [1]	Red Knot	Grey Plover
Whimbrel [4]	Sanderling	Ringed Plover [3]
Eastern Curlew [4]	Little Stint [3]	Little Ringed Plover
Common Redshank [2]	Red-necked Stint	Kentish Plover [3]
Marsh Sandpiper	Long-toed Stint [2]	Lesser Sand Plover [4]
Common Greenshank [4]	Baird's Sandpiper [3]	Greater Sand Plover [4]
Green Sandpiper [3]	Pectoral Sandpiper [2]	Caspian Plover [3]
Wood Sandpiper	Sharp-tailed Sandpiper	Oriental Plover [1]
Terek Sandpiper [4]	Curlew Sandpiper	Oriental Pratincole [1]

Table 2. Non-shorebird migrants from the Palearctic, Asia and New Guinea.

Garganey [2]	White-winged Black Tern	Dollarbird [5]
Eurasian Little Grebe [3]	Common Tern	Barn Swallow
Torresian Imperial-Pigeon [5, 6]	Oriental Cuckoo	Red-rumped Swallow [3]
Elegant Imperial Pigeon [3]	Common Koel [5]	Oriental Reed-Warbler [3]
Black-headed Gull [3]	Channel-billed Cuckoo [5]	Yellow Wagtail
Sabine's Gull [3]	Fork-tailed Swift	Grey Wagtail [2]
Black-tailed Gull [3]	House Swift [3]	

[1] Occurs Sep - Dec, on southward passage. Records outside this period are rare.
[2] Rare species (not recorded every year in the Top End).
[3] Vagrant.
[4] Frequently overwinters in the Top End.
[5] Breeds in the Top End.
[6] Present in small numbers throughout the year.

Dry Season Migrants

Relatively few species can be considered true Dry Season migrants, moving completely out of the Top End during the Wet Season (Table 3). However, for many species the majority of individuals move southward to breed during the Wet Season (the Austral summer). Some individuals move to inland and temperate coastal areas, while others move only relatively short distances.

Table 3. Dry Season migrants to the Top End. While some arrive during the latter part of the Wet Season, most arrive during April and May and some depart as late as December.

Straw-necked Ibis	Australian Pratincole	Black-eared Cuckoo
Glossy Ibis	Whiskered Tern	White-winged Triller
Black Kite	Pallid Cuckoo	Black-faced Cuckoo-shrike
Little Eagle	Cockatiel	Tree Martin
Swamp Harrier	Spotted Nightjar	Fairy Martin
Black Falcon	White-breasted Woodswallow	Welcome Swallow [3]
Red-necked Avocet	White-browed Woodswallow	
Red-kneed Dotterel	Little Woodswallow	

Nomads and irruptive species

The movements of some species into and within the Top End are irregular and largely unpredictable (Table 4). Nomadic species move around depending on the availability of food (e.g. nectar and grain) or water. The 'irruptive' species build up numbers in their main breeding area after a year or years of high breeding success. Harsh climatic conditions in the following season(s) cause them to disperse widely, appearing and sometimes breeding, where they may have been absent for many years.

Table 4. Nomads and irruptive species.

Emu [2]	Little Bittern [3]	Budgerigar [1]
Hoary-headed Grebe	Baillon's Crake	Singing Honeyeater [2]
Great Crested Grebe	Australian Crake	Grey-headed Honeyeater [2]
Black Swan	Spotless Crake	Grey-fronted Honeyeater [2]
Letter-winged Kite [1]	Lewin's Rail [3]	Black Honeyeater
Australian Wood Duck	Black-tailed Native-hen [1]	Crimson Chat [1]
Freckled Duck	Australian Bustard [2]	
Pink-eared Duck	Flock Bronzewing	
Eurasian Coot	Yellow-billed Spoonbill	

[1] Irruptive species.
[2] Small numbers resident in the Top End.
[3] Rare.

The Top End Climate

The Top End climate is characterised by quite distinct Wet and Dry seasons (Fig 1). The Wet Season is considered to be from October to April, though the only really wet months are from December to March, when about 80% of the annual rainfall occurs. The Dry Season, from May to September, typically has blue skies, warm dry days, low humidity and cool nights.

Temperatures (Fig 2) vary little throughout the year. During the Dry Season, the average maximum temperature in Darwin is about 31°C (22°C overnight), though June and July may be somewhat cooler. Temperatures over the inland Top End are similar during the daytime but significantly cooler at night.

From October to April, the average maximum in Darwin is 32°C (25°C overnight). Inland, daytime temperatures are higher, with overnight temperatures a little cooler than Darwin.

October and November are generally the hottest time of year, though these months are usually great for birdwatching. Many of the Dry Season migrants can still be seen and most of the Wet Season migrants have arrived, along with the few passage migrants that stay around for only a month or so before continuing southwards.

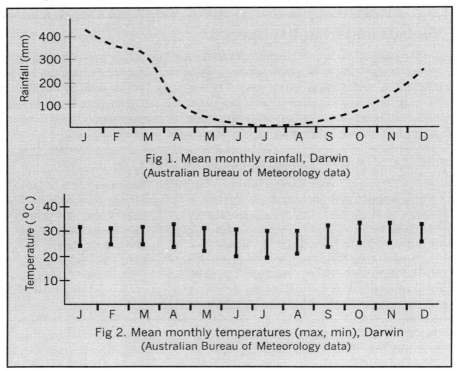

Fig 1. Mean monthly rainfall, Darwin
(Australian Bureau of Meteorology data)

Fig 2. Mean monthly temperatures (max, min), Darwin
(Australian Bureau of Meteorology data)

Practicalities

Getting around

Public transport in the Top End, even in Darwin, is of limited help in getting to most birding sites. To visit most of the sites in this book, having access to your own vehicle is really the only option.

Driving in the Top End

Overseas visitors can drive in the Northern Territory for up to three months on a Country of Origin' licence providing it is in English. Non-English licences require an International Driving Permit obtained from the country of origin.

The roads to most of the birding sites in this book are fully sealed. The few that are unsealed are specified on the maps and incidental road damage notwithstanding, are suitable for a standard vehicle driven with care.

Some of the roads in the Northern Territory have no speed limit. The distances can be quite great and the temptation is to cover these distances quickly. Most serious accidents in the Northern Territory are single vehicle accidents, very frequently the result of excessive speed or weariness. It is recommended to take a break from driving every two hours.

Between December and April some roads are affected by rain and it is advisable to check road conditions before departure. Information on Northern Territory road conditions can be obtained on the Internet or by telephone (see *Further Information*, p.160). In addition to this, reliable local advice can be obtained en route at Police Stations, information centres and fuel stops.

Accommodation

For most of the sites in the book there is access to hotel type accommodation close enough to make it a practical option. Camping grounds are well distributed in the National Parks and towns. They generally have good facilities and are inexpensive. For up to date information on a full range of accommodation options, check with the NT Tourist Commission (see *Further Information*, p.160).

How to use this book

1. Site guides
The brief introduction to each site describes its location, habitats and the birds of most interest. Where numbers of species recorded at a site or a region are given, these are the authors' own totals.

Key Species
These birds can be considered the highlights for the site because they are rare, restricted in range or difficult to find. Species that are particularly rare are included if there is a reasonable possibility of their occurrence. Vagrants are not listed here but may be mentioned in the site text, for informational purposes. Seasonality of occurrence is described in the Annotated List and is generally not repeated in the site descriptions.

Other Species
This is a sample of some of the more common birds seen at the site. Specific detail on finding them is not generally given.

Finding Birds
This section gives specific details on how to locate the birds, particularly the *key species*. Numbers in square brackets refer to corresponding numbers on the accompanying map. The amount of information provided on finding a particular bird is related to its rarity or to the level of difficulty in finding it.

Maps
The maps (including distance scales and north pointers) are accurate enough for birdwatching but are not meant for use as the only means of navigation in an area. Numbers on the maps are referred to in the accompanying text. Vegetation and topographical features are indicated if it is particularly useful. Where the distance to another locality is given on a map, this is measured from the point marked with the symbol ⊗. Grid references, where provided, are based on the WGS 84 Datum.

2. Annotated List of Birds of the Top End
All species recorded for the region, including confirmed vagrants, are listed with a summary of their distribution, abundance and seasonality of occurrence.

3. Regional Checklist
This is a tabulated list of species showing the regions in which each species occurs. Use it to help create an itinerary for a birding trip around the Top End. It will also indicate whether a species is seen out of its normal range, so that it may be documented and reported.

1 Darwin Region

Darwin, the capital of the Northern Territory, is the only city in the Top End that has a major airport, so for many visiting birdwatchers it will be the starting point of their Top End birding experience.

The regional birdlist totals approximately 310 species, which is higher than other regions in the Top End. Access to coastal and mangrove habitats is easier in the Darwin area than elsewhere in the Top End. There are some excellent wetlands just outside Darwin and these are particularly good for birdwatching from about June to December. Small reserves such as at Howard Springs and Holmes Jungle provide good areas of monsoon forest, where Rainbow Pitta, Rose-crowned Fruit-Dove and other monsoon forest birds can be seen. There are also quite extensive areas of open savanna woodland in the region but there are fewer bird species in this habitat in the Darwin region than in similar habitat further inland.

Darwin has an extensive range of accommodation options, including camping grounds, backpacker hostels and hotels (see *Further Information*, p.160).

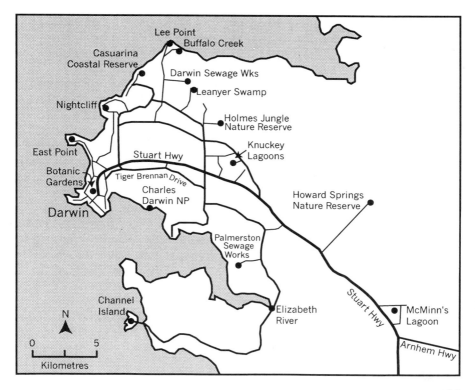

1.01 Darwin Harbour

There are a number of points around the city area providing good access to the shore and any of these can be productive, especially for herons, terns, gulls and shorebirds. Some access points will undoubtedly disappear as further commercial development occurs – take advantage of those that are still available.

Key Species
Lesser Frigatebird, Brown Booby, Eastern Reef Egret, Striated Heron, Beach Stone-Curlew, Azure Kingfisher, Collared Kingfisher, Cicadabird, Helmeted Friarbird

Other Species
Orange-footed Scrubfowl, Eastern Curlew, Grey-tailed Tattler, Terek Sandpiper, Little Tern, Common Tern, Brush Cuckoo, Australian Koel, White-breasted Woodswallow, Figbird

Finding Birds
Anywhere that allows access to the shore is worth checking. However, there are a few areas that are particularly worth visiting.

Fisherman's Wharf [1] provides good views of muddy shores at low tide, where Bar-tailed Godwit, Eastern Curlew, Whimbrel, Terek Sandpiper and Grey-tailed Tattler regularly occur in season. Eastern Reef Egret and Striated Heron are also frequently found here and Collared Kingfisher may be seen feeding on the open mud or perched around the edge of the shore.

Stokes Hill Wharf [2] is one of the most productive areas of the harbour and is the site where any vagrant gulls tend to be seen. Look out to sea from the wharf for terns and gulls, watching also for Lesser Frigatebird during the Wet Season. There are good views of muddy shore [3], where Beach Stone-Curlew can sometimes be seen in the early morning at low tide. Bar-tailed Godwit, Eastern Curlew, Whimbrel, Terek Sandpiper, Grey-tailed Tattler and Greater and Lesser Sand Plovers are also commonly seen on the mud here, along with Eastern Reef Egret and Striated Heron. Small numbers of White-breasted Woodswallow may be seen around this area throughout the year, whereas in the rest of the region it is only a Dry Season visitor.

Fort Hill Wharf [4-5] has limited access but still provides some good birding opportunities. Numerous gulls and terns roost on the loading ramps [5] and these can be seen from the road. It is possible to walk onto an unused part of the wharf [4] to gain views of more roosting areas as well as views of the muddy shore. As the tide recedes, many gulls and terns feed along the open shore [6]. Eastern Reef Egret and Striated Heron are common here and from October to April, Eastern Curlew, Whimbrel, Terek Sandpiper and Grey-tailed Tattler regularly occur here. Brown Booby and Lesser Frigatebird may also be seen here at this time of year. Park off the road [P1] and walk to the top of the low cliff to gain views.

The Esplanade runs alongside an extensive parkland [8] where Helmeted Friarbird, Rufous-banded Honeyeater and Figbird are commonly seen. The adjacent shoreline [7] is a low tide feeding area for gulls, terns and shorebirds.

Doctor's Gully [see inset] is an expansive mudflat at low tide, with a patch of monsoon forest along a small stream that runs into the sea. Park in the main carpark [P2] and check the close shoreline [9] for Beach Stone-curlew, Striated Heron and Eastern Reef Egret. Further out, if there are large areas of mud, watch for terns and shorebirds. Australian Koel, Brush Cuckoo, Helmeted Friarbird and Figbird frequently occur in the lush growth on the hillside [10] at the edge of the carpark. From the smaller carpark [P3] a boardwalk follows the small stream through the forest [11]. Azure Kingfisher occurs here and can often be seen perched on the rails of the boardwalk. Orange-footed Scrubfowl is regularly seen on the slopes at the sides of the stream and Cicada-bird is sometimes seen in this area.

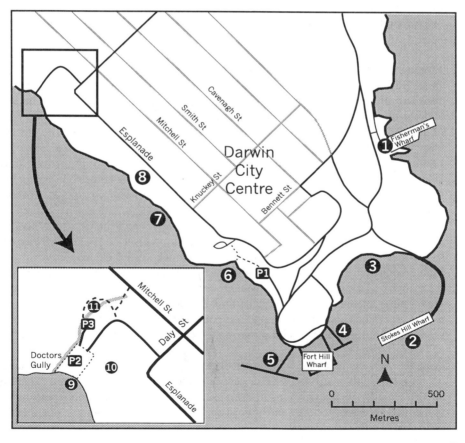

1.02 Tiger Brennan Drive

Tiger Brennan Drive provides easy access to mangroves very close to Darwin city and has a good range of mangrove birds. Along Tiger Brennan Drive, north of the suburb of Bayview Haven, there are a few short tracks leading to electricity pylons in and adjacent to the mangroves. Any of these areas can be productive but birding around Bayview Haven itself has been the most successful over many years. As with other mangrove areas, the height of the tide is critical when looking for Chestnut Rail. Aim to get there when the tide is falling just below 4 m and ideally when that occurs in the morning. Nevertheless, Chestnut Rail may be seen here at any time of the day, if there is exposed mud to provide a feeding area for the birds.

Key Species

Great-billed Heron, Grey Goshawk, Chestnut Rail, Collared Kingfisher, Green-backed Gerygone, Red-headed Honeyeater, Mangrove Robin, Mangrove Grey Fantail, Black Butcherbird, Yellow White-eye

Other Species

Osprey, Brahminy Kite, Red-winged Parrot, Helmeted Friarbird, Rufous-banded Honeyeater

Finding Birds

Enter Bayview Haven via Stoddart Drive, at the traffic lights. The footpath beside Stoddart Drive [1] provides a good view over the mangroves. Yellow White-eye, Helmeted Friarbird, Mangrove Robin and Black Butcherbird may be seen at the edge of the mangroves here, directly by the footpath. Osprey is frequently seen nearby and Grey Goshawk is occasionally seen flying over or perched atop taller mangroves.

Follow the road or footpath to the lock [3], where there is a view of the estuary and a chance of seeing Great-billed Heron. Chestnut Rail regularly comes out during low tide to feed at the mangrove edge below the footpath between [3] and [2]. Collared Kingfisher also occurs here.

A short track about 400m along Tiger Brennan Drive towards the city from Stoddart Drive provides access to another good area of mangroves [see inset]. Park off the road by the gate to the access track [4] and follow the track toward the mangroves.

Mangrove Robin is frequently seen in the area around the concrete block [5], where the mangroves are relatively short. Mangrove Grey Fantail occurs in the denser mangroves but is uncommon. Black Butcherbird, Helmeted Friarbird, Red-headed Honeyeater, Large-billed and Green-backed Gerygones and Yellow White-eye are commonly seen here. At times these species move to the scrubbier bush [8] on the opposite side of Tiger Brennan Drive.

From the concrete block [5], a muddy track goes deeper into the mangroves to a large opening [6], with small pools or streams of water and expanses of mud. There is another, smaller, track branching off to a similar spot [7]. Take care along these tracks, as the mud here is soft, sticky and quite deep. Chestnut Rail may venture out onto the open mud at either of [6] or [7].

The carpark at the end of Gonzalez Road [9] provides a good vantage point to view the mangrove edge from the seaward side. Great-billed Heron is sometimes seen here. Striated Heron and Eastern Reef Egret are regularly present and, from September to April, Terek Sandpiper, Eastern Curlew, Greater Sand Plover and other shorebirds feed on the open mud.

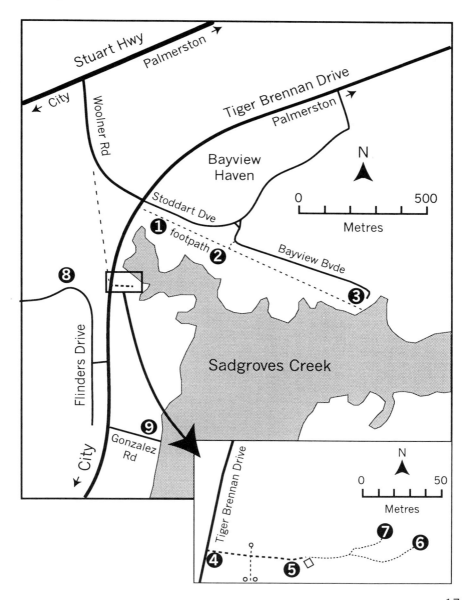

1.03 Charles Darwin National Park

The entry road to the Park is from Tiger Brennan Drive, about 6 km from Darwin city. This small park has quite an extensive area of woodland and mangroves, though access to the mangroves is limited at high tide. Heed the signs that warn of biting insects by covering up well and using an effective repellent. The park is open between 7 am and 7 pm daily.

Key Species
Brown Quail, Chestnut Rail, Bush-hen, Northern Rosella, Green-backed Gerygone, Mangrove Gerygone, Helmeted Friarbird, Silver-crowned Friarbird, Red-headed Honeyeater, Bar-breasted Honeyeater, Mangrove Robin, Grey Whistler, Broad-billed Flycatcher, Grey Butcherbird, Black Butcherbird

Other Species
Brahminy Kite, White-bellied Sea-Eagle, Red-winged Parrot, Rainbow Bee-eater, Little Friarbird, White-gaped Honeyeater, White-throated Honeyeater, Leaden Flycatcher, White-bellied Cuckoo-shrike, Varied Triller, White-winged Triller

Finding Birds
Brown Quail is often seen crossing the entry road [1], particularly early in the morning or late in the afternoon. The woodland along the entry road [1] can be good for birding, though many of the typical woodland birds, such as Red-winged Parrot and Silver-crowned Friarbird, can be seen just as easily by the carpark [P], or around the edges of the grassed picnic area [2]. Watch from the viewing platform at the edge of the picnic area for Osprey, White-bellied Sea-Eagle and Brahminy Kite, which can often be seen soaring over the harbour.

For what is generally the best birding, take the steps [3] down to the lower track and turn left then right, through the open woodland [4]. Here Northern Rosella, White-throated Honeyeater, Little Friarbird, Silver-crowned Friarbird, Long-tailed Finch, White-winged Triller and Grey Butcherbird can be seen.

The short track into the mangroves [5] can be muddy depending on how recent and high the previous tide has been. It is well worth taking this track as Mangrove Robin, Grey Whistler, Green-backed and Mangrove Gerygones, Yellow White-eye and Red-headed Honeyeater are all quite common. Broad-billed and Leaden Flycatchers are both found in the mangroves here, so care is needed with identification. Helmeted Friarbird is common in the mangroves and is occasionally also seen in surrounding woodland. Chestnut Rail occurs in these mangroves but is usually heard rather than seen. While it is possible to track the calling birds, it is difficult venturing into the mangroves where there are no tracks. The best option is to sit patiently and hope that a Rail may venture near enough to get views.

Bush-hen has been seen in tall grass at the edge of mangroves [6] during the Wet Season. It is possible that this species is here during the Dry Season but there have been no confirmed records.

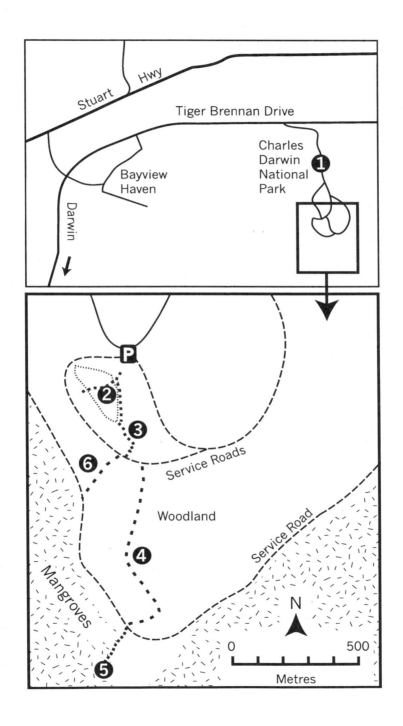

1.04 Darwin Botanical Gardens

The Botanical Gardens are close to the city and have a birdlist of about 50 species. The Gardens are an easily accessible and generally reliable site for Rufous Owl, which can frequently be seen roosting during the daytime. There are two access points to the Gardens - the 'top' entrance, from the Stuart Highway (using carpark P2) and the 'bottom' entrance, from Gardens Road (using carpark P1). The bottom entrance is most convenient for birdwatching, so directions are given from there. The months given for where to find Rufous Owl are approximate and may vary between years.

Key Species

Little Curlew, Brush Cuckoo, Channel-billed Cuckoo, Rufous Owl, Barking Owl, Rose-crowned Fruit-Dove, Green-backed Gerygone

Other Species

Orange-footed Scrubfowl, Magpie Goose, Glossy Ibis, Straw-necked Ibis, Australian White Ibis, Brahminy Kite, Torresian Imperial-Pigeon, Australian Koel, Forest Kingfisher, White-gaped Honeyeater, Dusky Honeyeater, Northern Fantail, Yellow Oriole, Figbird

Finding Birds

The Rufous Owls roost in one of two areas depending on the time of year (see below). In both areas it is a matter of checking suitable trees - these are typically open from below, with a fairly broad and dense canopy above. The owls show little or no evidence of trying to hide, usually perching away from the main trunk in clear view from below, though of course they can still be easily overlooked.

Lower area (Feb - Aug)

It is usually easier to find the owls here, as there are relatively few trees to search. The owls often roost in trees directly beside the toilet block [2], so try here first. Otherwise, it is a matter of checking all of the trees nearby, including those on the other side of the Amphitheatre access road [9], or a little way up the slope near the play equipment [3].

Upper Rainforest area (Sep - Jan)

The owls are often in a tree directly at the bottom of the small downward path [6]. Alternatively, there are a few favoured trees within about ten metres either side of this. Another favourite roosting area is by the small pond [7]. If the birds are not at any of these spots, there is no option but to wander around the rainforest tracks checking likely trees as described above. It can also be worth checking with the gardens' staff, as they may know where the owls are currently roosting. On occasion the presence of a Rufous Owl can be given away by the calls and behaviour of White-gaped Honeyeaters - listen for the excited alarm calls and there is a good chance that an owl will be nearby.

Barking Owl sometimes roosts in the trees by the fountain [1] or, more frequently, in trees along the path [4] between the fountain and the rainforest area. For this species it is best to visit just prior to dusk when birds start to call and may be more easily located. Channel-billed and Brush Cuckoos, Australian Koel and Rose-crowned Fruit-Dove are sometimes seen in the trees at [4].

Green-backed Gerygone is frequently found in trees near the toilet block [2] or further up the hill [3] toward the upper car park. Dusky Honeyeater may also be seen here.

Little Curlew often feed on the open grassy areas [5], along with Magpie Goose, Australian White Ibis, Straw-necked Ibis and occasionally Glossy Ibis.

Orange-footed Scrubfowl occurs throughout the gardens, most commonly in the rain-forest area [6-7] but often in the open areas around the lower toilet block [2].

Forest Kingfisher is frequently present in the large trees [8] near the bottom carpark.

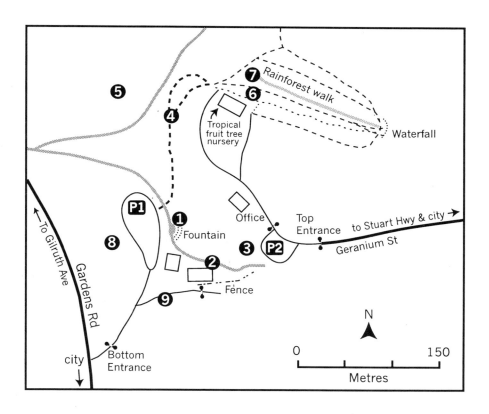

1.05 East Point Reserve

East Point Reserve is about 5 km or little more than five minutes drive from the city. It has a combination of monsoon forest, mangroves, reef and mudflats that provide habitat for a wide range of birds. It can be busy at weekends though disturbance to the birds is not usually a problem. The boardwalk in the mangroves, though short, can be productive. A good range of shorebirds and terns can be seen on the reef and mudflats when the tide is high enough to concentrate them; however, when it is too high the birds are forced to move elsewhere.

Key Species
Brown Booby, Lesser Frigatebird, Eastern Reef Egret, Common Redshank, Asian Dowitcher, Swinhoe's Snipe, Beach Stone-curlew, Bush Stone-curlew, Oriental Plover, Lesser Crested Tern, Rose-crowned Fruit-Dove, Oriental Cuckoo, Large-tailed Nightjar, Australian Owlet-nightjar, Collared Kingfisher, Red-backed Kingfisher, Rainbow Pitta, Mangrove Gerygone, Green-backed Gerygone, Broad-billed Flycatcher, Mangrove Grey Fantail

Other Species
Orange-footed Scrubfowl, Striated Heron, Eastern Curlew, Great Knot, Bar-tailed Godwit, Crested Tern, Rainbow Bee-eater, Red-headed Honeyeater, Grey Whistler, Lemon-bellied Flycatcher, Northern Fantail, Varied Triller, White-winged Triller, Yellow Oriole, Yellow White-eye

Finding Birds
Mangrove Boardwalk
Park at the Lake Alexander carpark [P1] and go through the gate, which has a sign for the boardwalk. The access track passes through an open grassy area, where Swinhoe's Snipe may be seen during the Wet Season. Lemon-bellied Flycatcher, White-winged Triller and Northern Fantail are often seen around the woodland edge. Follow the track through some denser woodland before getting into the mangroves and finally onto the mangrove boardwalk. Mangrove Grey Fantail, Broad-billed Flycatcher, Mangrove Gerygone, Red-headed Honeyeater and Yellow White-eye may be seen here.

Monsoon Forest
Park in the marked 'Barbecue Area' carpark [P2], just beyond the Peewee Restaurant, which is signposted. The track [1] on the opposite side of the road passes through a clearing in the monsoon forest. A short distance along this track there are smaller tracks to the left and right, marked with walking signs [1]. The left hand track is generally more productive, with quite regular sightings of Rainbow Pitta and Large-tailed Nightjar as well as occasional records of Australian Owlet-nightjar. Grey Whistler and Orange-footed Scrubfowl are quite common in this part of the forest. Rose-crowned Fruit-Dove may be seen anywhere in the forest here but is most frequently found toward the end of the main track [1].

Bush Stone-curlew is often seen resting under trees or in the open grassy area on the western side of the road near [2]. Oriental Cuckoo may be seen around the small patch of dense forest at the edge of the equestrian centre [3].

Foreshore

Red-backed Kingfisher often perches on power lines around the carpark [P3]. Oriental Cuckoo, Varied Triller, Green-backed Gerygone and a range of honeyeaters may be seen in the trees here. Many shorebirds including Great Knot, Lesser and Greater Sand Plovers, Terek Sandpiper and Grey-tailed Tattler roost on the reef [4] at high tide. Beach Stone-curlew often roosts here and even on lower tides there is a chance of seeing one. The reef is also a good place for Crested, Lesser Crested and Gull-billed Terns, Eastern Reef Egret and Striated Heron. Brown Booby and Lesser Frigatebird are often seen off shore from October to April.

The section of reef at [5] also serves as a shorebird roost though numbers of birds are generally lower here. Common Redshank and Asian Dowitcher have been seen here, though both are rare. Rose-crowned Fruit-Dove and Yellow Oriole may be seen in the large trees by the picnic shelter [P4] and Green-backed Gerygone occurs in the trees along the edge of the beach. Beach Stone-Curlew may be seen here or further along the shore towards the mangroves [6].

If the tide is very high, birds may be forced out of this area, so try the mudflats near the end of Colivas Road [7]. Eastern Curlew, Great Knot and Bar-tailed Godwit are common at high tide, and Beach Stone-curlew sometimes also roosts here at high tide. Finches and mannikins feed in the grass near the end of Colivas Road.

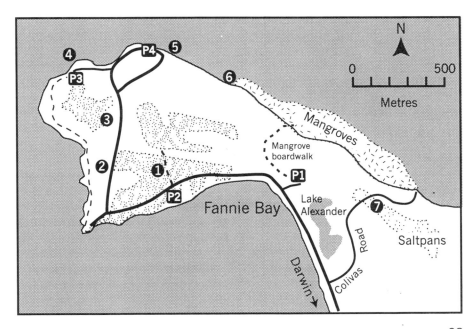

1.06 Nightcliff Foreshore

Nightcliff is a coastal suburb about 11 km north of Darwin city. Approximately 80 species have been recorded on and near the foreshore here. This includes many reef and mangrove birds, with a few records of the scarce Mangrove Golden and White-breasted Whistlers. From September to April a large range of shorebirds use the reef as a high tide roost. Frigatebirds and boobies may also be seen inshore here, particularly during windy weather.

Key Species
Brown Booby, Lesser Frigatebird, Striated Heron, Eastern Reef Egret, Pacific Baza, Oriental Plover, Beach Stone-curlew, Sooty Oystercatcher, Pied Oystercatcher, Collared Kingfisher, Barking Owl, Tawny Frogmouth, Helmeted Friarbird, Red-headed Honeyeater, White-breasted Whistler, Mangrove Golden Whistler, Yellow White-eye

Other Species
Pied Cormorant, Grey-tailed Tattler, Terek Sandpiper, Common Sandpiper, Great Knot, Sanderling, Pacific Golden Plover, Greater Sand Plover, Lesser Sand Plover, Peaceful Dove, Bar-shouldered Dove, Little Friarbird, Brown Honeyeater, Rufous-banded Honeyeater, Magpie-lark

Finding Birds
Brown Booby and Lesser Frigatebird are occasionally seen close inshore from October to April. Shorebirds, including Oriental Plover, Greater and Lesser Sand Plovers, Great Knot, Grey-tailed Tattler, Terek Sandpiper, Pied Oystercatcher and sometimes Beach Stone-curlew roost on the reef [1-4] at high tide.

Collared and Sacred Kingfishers, Eastern Reef Egret and Striated Heron may be seen on the reef when the tide is low. Sooty Oystercatcher is sometimes seen on the reef also, though more often it can be seen roosting on the breakwater [2] near the jetty. Yellow White-eye, Helmeted Friarbird and Rufous-banded, Red-headed and Brown Honeyeaters regularly occur in the small mangroves along the edge of the shore near the reef [1-4]. Mangrove Golden and White-breasted Whistlers have been seen, albeit rarely, in the larger and denser mangroves along the shore south of [4]. When the tide is low it is possible to walk along the seaward side of the mangroves to look for these species. Avoid being caught on an incoming tide by doing this walk as the tide is receding. Tidal information may be found in the local newspaper or on the Internet (see *Further Information*, p.160).

Tawny Frogmouth is sometimes seen in the pine-like *Casuarina* trees near the car park [3]. At night Barking Owl often hunts by the street lights here or around the carparks [P1, P2]. The footpath between [P1] and [P2] passes through open parkland and Little Friarbird, Blue-faced and Dusky Honeyeaters and Double-barred Finch can be seen here. Watch here and the adjacent residential area for Pacific Baza, as it is known to nest in the large trees around Nightcliff.

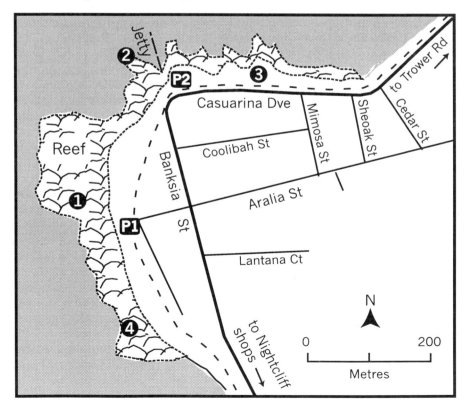

1.07 Casuarina Coastal Reserve

Casuarina Coastal Reserve is at the edge of Darwin's northern suburbs. It is an extensive area covering a range of habitats including beach, monsoon forest, grassland and some woodland, accessed via a number of paths and boardwalks. Over 100 species have been recorded here, with a number of mangrove birds quite easy to see. The tracks and mangrove boardwalks can become inundated or boggy during the Wet Season or at high tide. There are a number of access points but the most convenient is adjacent to the Darwin Hospital, in the northern suburb of Tiwi.

Key Species

Grey Goshawk, Bush-hen, Red-backed Button-quail, Rose-crowned Fruit-Dove, Emerald Dove, Little Bronze-Cuckoo, Barking Owl, Little Kingfisher, Azure Kingfisher, Mangrove Gerygone, Green-backed Gerygone, Black Butcherbird, Helmeted Friarbird, Rufous-throated Honeyeater, Cicadabird, Yellow White-eye

Other Species

Brown Quail, Orange-footed Scrubfowl, Nankeen Night Heron, White-bellied Sea-Eagle, Pheasant Coucal, Tawny Frogmouth, Rainbow Bee-eater, Dollarbird, Red-backed Fairy-wren, Grey Whistler, Weebill, Rufous-banded Honeyeater, Red-headed Honeyeater, Lemon-bellied Flycatcher, Shining Flycatcher, White-winged Triller, Long-tailed Finch, Crimson Finch, Double-barred Finch, Chestnut-breasted Mannikin

Finding Birds

Park in either of the marked carparks [P] along Paracelsus Road and walk down the short slope to the track. At the bridge over the small creek [1] watch for finches and honeyeaters coming in to drink during the Dry Season, particularly in the late afternoon. Brown, Rufous-banded and White-throated Honeyeaters are regular but in some years Rufous-throated Honeyeater also occurs here. White-winged Triller may be seen in the drier vegetation along track [10]. Weebill occurs in the woodland [1-2]. At the woodland edge [2] look for Brown Quail, Dollarbird, Barking Owl and Tawny Frogmouth. Bush-hen sometimes occurs in the grassy area between [2] and [3] during the Wet Season but can be very difficult to locate. Listen for its call in the early morning or late afternoon. With patience it is possible to track down the calling bird. Mangrove, Large-billed and Green-backed Gerygones, Red-headed Honeyeater, Lemon-bellied Flycatcher and Yellow White-eye breed around the open saltpan in the mangroves [3] and are quite common. Black Butcherbird, Grey Whistler, Cicadabird, Shining Flycatcher and Little Bronze-Cuckoo also occur here. Striated Heron is occasionally pushed onto this saltpan by the rising tide. Emerald Dove and Rose-crowned Fruit-Dove may be seen throughout the mangroves and adjoining woodland, though are most often seen further into the mangroves [4], as is Nankeen Night Heron. Little Kingfisher can sometimes be seen at the small creek at the end of the mangrove track [5], or, when the tide is low, at the creek near the boardwalk [8]. Crimson, Double-barred and Long-tailed Finches and Chestnut-breasted Mannikin often feed in the grass along the edge of track [6] and Little Bronze-Cuckoo and Pheasant Coucal are frequently seen here also. Look for Red-backed Button-quail in the grassy area around the track junction [7]. Red-backed Fairy-wren and Leaden Flycatcher are often present in the open woodland [9].

Osprey, White-bellied Sea-Eagle and Brahminy Kite frequently occur here and Grey Goshawk is an occasional visitor to both the mangrove and woodland areas.

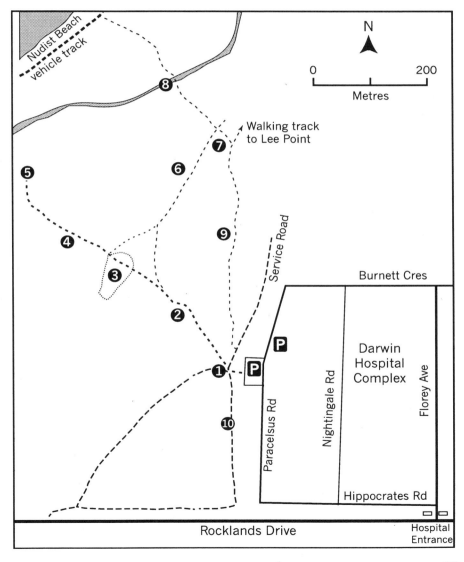

1.08 Lee Point

Lee Point is the northernmost tip of the Darwin area, about 18 km from the city. It is one of the few places in the Darwin area that face open sea and hence most of the seabird records are from here. The shore is particularly good for terns and is also a feeding and roosting area for large numbers of shorebirds. At weekends and on most afternoons, people with dogs disturb the birds, so the shore is best visited during the week, in the morning. There is, at the time of writing, a move to more strongly enforce the dog exclusion zone between Lee Point and Buffalo Creek in order to protect the shorebirds.

Key Species
Brown Booby, Lesser Frigatebird, Grey Goshawk, Pacific Baza, Black Falcon, Little Curlew, Oriental Plover, Sooty Oystercatcher, Lesser Crested Tern, Little Tern, Rose-crowned Fruit-Dove, Varied Lorikeet, Tawny Frogmouth, Green-backed Gerygone, Helmeted Friarbird, Dusky Honeyeater, Red-headed Honeyeater, Barn Swallow

Other Species
Brown Goshawk, Australian Hobby, Bar-tailed Godwit, Whimbrel, Eastern Curlew, Common Greenshank, Grey-tailed Tattler, Ruddy Turnstone, Great Knot, Red Knot, Sanderling, Red-necked Stint, Grey Plover, Greater Sand Plover, Lesser Sand Plover, Common Tern, Crested Tern, Red-tailed Black-Cockatoo, Red-winged Parrot, Pheasant Coucal, Rainbow Bee-eater, Striated Pardalote, Silver-crowned Friarbird, White-gaped Honeyeater, Northern Fantail, Spangled Drongo, Varied Triller, Yellow Oriole, Olive-backed Oriole, Great Bowerbird, Double-barred Finch, Masked Finch, Long-tailed Finch, Chestnut-breasted Mannikin

Finding Birds
Grey Goshawk and Pacific Baza can sometimes be seen in the woodland around the corner of Fitzmaurice Drive [1]. Long-tailed, Crimson and Double-barred Finches are often present in the grassy areas by the fence here. Striated Pardalote occurs regularly here in the Dry Season, and is frequently seen perched on power lines beside Lee Point Road.

From September to November Oriental Plover and Little Curlew may be seen in the extensive grasslands [2]. However, this is a restricted area, so viewing is only possible from the roadside. Raptors that have been seen hunting over the grasslands include Nankeen Kestrel, Brown Falcon, Black Falcon, Black Kite and Brown Goshawk.

A narrow bitumen road running alongside the caravan park provides excellent access to the woodland and to a small reservoir [4]. Silver-crowned Friarbird and Great Bower-bird are regularly seen in the woodland. Varied Lorikeet may occur in large groups here when trees are flowering. Long-tailed Finch feeds in the grass along the caravan park fence line. The reservoir rarely holds waterfowl but late in the Dry Season (September and October) a range of honeyeater species and Crimson, Long-tailed and Masked Finches drink here.

Red-tailed Black-Cockatoo occurs regularly in the woodland [3] opposite the caravan park. Varied Lorikeet, Red-winged Parrot and Silver-crowned Friarbird are frequently seen here also and Long-tailed Finch often perches on the nearby electric power lines. Many of the woodland birds come to the caravan park during the day to drink at the taps and other water sources.

At the Point itself, Varied Triller, Leaden Flycatcher and Spangled Drongo are often seen near the picnic areas near the car park [P]. Behind the toilet block [5] there are tracks giving access to the woodland where these species and other common woodland birds may be seen. Pacific Baza has been seen in the taller trees at the edges of the grassy picnic areas.

At the point itself, Lesser Crested Tern frequently roosts around the small reef area [6] and Little Tern roosts nearby on the sand. Pied Oystercatcher and sometimes Sooty Oystercatcher can be seen on the reef or together with other shorebirds at high tide roosts on the sand. Large numbers of Sand Plovers (mainly Greater) roost on the beach on the eastern side of the point [7]. Lesser Frigatebird, Brown Booby and Barn Swallow are often seen flying quite close inshore from October to March, particularly during windy weather.

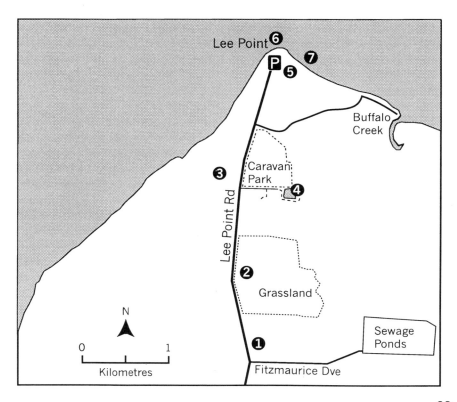

1.09 Buffalo Creek

Buffalo Creek is 18 km north of the city, near Lee Point (site 1.08). About 150 species have been recorded here, including several national rarities. Birdwatching is good throughout the year at this site, as there is an excellent range of shore, mangrove and monsoon forest species. It is best early in the morning, on a weekday, as motorboat launching and dogs can be disruptive particularly at weekends. Tide levels are critical for Chestnut Rail and to a lesser extent for shorebirds, so check the local newspaper, *NT News*, for tide times and heights.

Key Species
Streaked Shearwater, Brown Booby, Lesser Frigatebird, Great-billed Heron, Chestnut Rail, Beach Stone-curlew, Sooty Oystercatcher, Asian Dowitcher, Lesser Crested Tern, Emerald Dove, Rose-crowned Fruit-Dove, Little Bronze-Cuckoo, Large-tailed Nightjar, Azure Kingfisher, Little Kingfisher, Rainbow Pitta, Green-backed Gerygone, Mangrove Gerygone, Large-billed Gerygone, Red-headed Honeyeater, Mangrove Robin, Mangrove Golden Whistler, White-breasted Whistler, Grey Whistler, Mangrove Grey Fantail, Black Butcherbird, Yellow White-eye

Other Species
Orange-footed Scrubfowl, White-bellied Sea-Eagle, Brahminy Kite, Great Knot, Bar-tailed Godwit, Black-tailed Godwit, Grey-tailed Tattler, Greater Sand Plover, Lesser Sand Plover, Gull-billed Tern, Caspian Tern, White-gaped Honeyeater, Dusky Honeyeater, White-throated Honeyeater, Lemon-bellied Flycatcher, Double-barred Finch

Finding Birds
Chestnut Rail can often be seen from the boat ramp [1], particularly on the opposite side of the creek to the right [2]. The best time to see the Rail is early in the morning, with the tide at 4m and falling. A reasonable amount of mud is then exposed and is shaded from bright sunlight. Unfortunately ideal times only occur for a few days each month; however, birds can still be seen coming out of the mangroves at any time of day if the tide conditions are suitable and there is no disturbance. There is a small, ill-defined track that fishers use to get access further along the creek [3]. This provides views of an extensive mangrove area and can be a good option but the track can be rather muddy and the makeshift 'bridge' over the small creek is difficult to cross.

Great-billed Heron is sometimes seen on the muddy banks of the creek, mainly early in the morning or very late in the afternoon. Azure Kingfisher is often present in the mangroves at the sides of the boat ramp [1] and Little Kingfisher and Mangrove Grey Fantail are sometimes also seen here.

Tide heights between about 5.5m and 6.5m are ideal to see shorebirds and terns roosting, as the water is high enough to concentrate the birds here but not high enough to inundate the area completely, at which time the birds fly to other roosting areas. Large numbers of Great Knot and Bar-tailed Godwit roost near the mouth of the creek [5]. Scan these flocks carefully for Asian Dowitcher, which occasionally occurs here. Greater and Lesser Sand Plovers usually roost a little further along the beach [6], which is also the most likely spot for Sooty Oystercatcher and Beach Stone-curlew. Other shorebirds such as Red-capped Plover, Ruddy Turnstone, Common Green-shank, Grey-tailed Tattler, Terek Sandpiper and Sanderling occur in small numbers along the tideline. Watch out to sea for Brown Booby and Lesser Frigatebird, particularly between October and April. There are very few records of true seabirds though Streaked Shearwater is a possibility from May to October.

Green-backed Gerygone, Red-headed Honeyeater and Yellow White-eye are normally easy to find in the mangroves at the edge of the main carpark [P]. Green-backed Gerygone may be seen throughout the monsoon forest areas, including in trees along the seaward edge of the monsoon forest [4], where it often feeds early in the morning. Rainbow Pitta, Grey Whistler, Black Butcherbird, Emerald Dove and Rose-crowned Fruit-Dove also occur in the monsoon forest. They are often found along the main entry road, or on the connecting tracks between that road and the beach [7, 8], particularly in the early morning. From the main carpark a track [9] passes through the monsoon forest/mangrove ecotone, where Little Bronze-Cuckoo and Mangrove Golden Whistler may be found, along with the more common mangrove birds. Large-tailed Nightjar is resident in the Buffalo Creek area. It is often seen at dusk, flying around the carpark [P] or around the open area by the bitumen road [10].

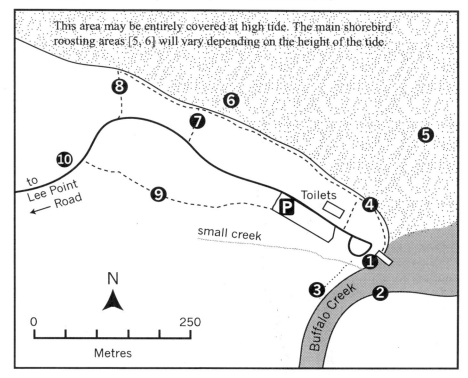

1.10 Leanyer Sewage Works

The Leanyer Sewage Works are about 16 km north of Darwin city, near the suburb of Leanyer. Access is via Fitzmaurice Drive, which runs off Lee Point Road. Well over 200 species have been seen within and immediately around the sewage works, including a number of national rarities. The sewage works consist of a series of open ponds surrounded by gravel and bitumen tracks. The area is not extensive and simply driving around can be productive. However, knowledge of the good spots will avoid some of the frustration of missing target species that may be disturbed if large numbers of waterfowl are put to flight. Waterfowl are generally plentiful from May to November, decreasing in number as the rains set in during December. The surrounding mangroves can be very productive, with birds often coming to the mangrove edge, where they can be seen from a vehicle.

Permission to enter (and a key where required) can be obtained from the Power & Water Authority provided reasonable notice is given. For contact details, see *Further Information* (page 160).

Key Species

Pink-eared Duck, Garganey, Bush-hen, Buff-banded Rail, Little Curlew, Oriental Plover, Little Ringed Plover, Oriental Pratincole, White-winged Black Tern, Common Tern, Flock Bronzewing, Little Bronze-Cuckoo, Mangrove Gerygone, Red-headed Honey-eater, White-breasted Whistler, Mangrove Golden Whistler, Grey Wagtail, Yellow Wagtail, Yellow-rumped Mannikin, Barn Swallow, Yellow White-eye

Other Species

Brown Quail, Australasian Grebe, Radjah Shelduck, Wandering Whistling-duck, Pacific Black Duck, Grey Teal, Hardhead, Pied Heron, White-bellied Sea-Eagle, Brahminy Kite, Australian Pratincole, Whiskered Tern, Crimson Finch, Long-tailed Finch, Chestnut-breasted Mannikin, Golden-headed Cisticola

Finding Birds

Many birds feed in and around the drain [1] immediately in front of the main gate when there is grass seeding. Crimson, Long-tailed and Double-barred Finches, Chestnut-breasted and (rarely) Yellow-rumped Mannikins, along with Brown Quail can be seen. Late in the Dry Season, if there is water in the drain, many birds come to drink here. Golden-headed Cisticola and Restless Flycatcher are frequently seen in or around the palm-like *Pandanus* trees at the side of the road [1, 3]. Bush-hen has been seen during the Wet Season in the woodland by the track [4] that runs from Fitzmaurice Drive toward Leanyer Swamp (see site 1.11).

Beside the sewage works a narrow track runs along the stream from [1] to [2]. Grey Wagtail has been seen here but more commonly it is a good spot for Yellow Wagtail. Brahminy Kite, Pied Heron and, from October - May, White-winged Black Tern and Common Tern are frequently seen around the ponds at the start of track [5].

There is an extensive area of mangroves adjacent to the sewage works and looking through the fence at the mangrove edges can be rewarding. Little Bronze-Cuckoo is frequently seen in mangroves along the edge of track [6]. Mangrove Gerygone, Yellow White-eye and Red-headed Honeyeater are often seen in mangroves along the eastern end of track [8]. Mangrove Golden and White-breasted Whistlers have also been recorded here in the past, so it is worth watching for them.

Wandering Whistling Duck (and less commonly, Plumed Whistling Duck), Pink-eared Duck, Hardhead, Radjah Shelduck and other waterfowl are most numerous along tracks [10] and [11] and in the adjacent ponds. Garganey has been recorded less in recent years but most records are from the ponds on either side of track [14]. Little Ringed Plover is most often found along tracks [9, 10, 11]. It is occasionally seen along track [5] or by rainwater puddles at the eastern end of track [8]. Yellow Wagtail may be seen throughout the sewage works but it is more likely along tracks [9] or [6] early in the season (Sep - Oct) and around track [14] late in the season (Mar - Apr). Little Curlew may be seen on almost any of the tracks, as can the much less common Oriental Plover and Oriental Pratincole. In recent years Flock Bronzewing has been seen quite frequently along tracks [10] and [11] from May to October. Check carefully, since even if one is present it can be difficult to pick out amongst the many hundreds of waterfowl along the tracks.

Buff-banded Rail is sometimes present if there is vegetation in the corner of the ponds, particularly in the pond at the southwestern end of track [9].

Check the power lines at the western end of track [8] for swallows. Barn, Welcome and Red-rumped Swallows have all been seen here, though the latter is a very rare vagrant to the Top End.

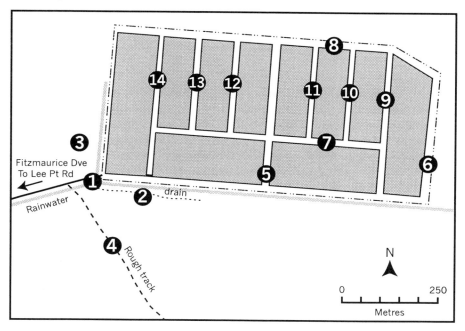

1.11 Leanyer Swamp

Leanyer Swamp is an extensive area with three fairly distinct sections. One is not publicly accessible. Another is accessed from Holmes Jungle Nature Reserve (site 1.12) and so it is treated in that section. The third, which at some point may be lost to development, is dealt with here. Artificial drains keep the area dry for much of the year but during most of the Wet Season there is generally some surface water that can provide suitable habitat for birds. Additionally, the larger high tides (greater than 6.5 m) can push shorebirds into the mudflat areas. The swamp is bordered along one edge by mangroves and dissected by small mangrove-lined creeks and consequently a few of the commoner mangrove birds can be seen.

There are two main access points to this part of Leanyer Swamp. From Fitzmaurice Drive, beside the sewage works [1], a rough track passes through woodland to the swamp area. This track is unsuitable for conventional vehicle at any time of year. From Hodgson Drive in the suburb of Leanyer, a shorter, smoother dirt track [9] connects to the main track that leads to the swamp. Whilst this track may be suitable for conventional vehicle in dry weather, it is likely that the entry gate will be locked. There is pedestrian access to the swamp from this gate.

Key Species

Black Bittern, Bush-hen, Buff-banded Rail, Spotted Harrier, Black Falcon, Swinhoe's Snipe, Broad-billed Sandpiper, Ruff, Long-toed Stint, Little Ringed Plover, Flock Bronze-wing, Varied Lorikeet, Yellow Wagtail, Yellow-rumped Mannikin, Zitting Cisticola

Other Species

Brown Quail, Radjah Shelduck, Black-necked Stork, White-bellied Sea-Eagle, Swamp Harrier, Black-shouldered Kite, Australian Pratincole, Little Curlew, Red-capped Plover, Pheasant Coucal, White-throated Honeyeater, Golden-headed Cisticola, Crimson Finch, Long-tailed Finch, Chestnut-breasted Mannikin

Finding Birds

There are extensive areas of short open grassland at this site, so species attracted to that habitat may be found throughout much of the area. Australian Pratincole is generally common here and Black-shouldered Kite, Spotted Harrier, Swamp Harrier and Black Falcon are sometimes seen hunting over the grassland.

Crimson and Long-tailed Finches along with Chestnut-breasted and (rarely) Yellow-rumped Mannikin may be seen in the vegetation along the edges of the creek line in front of the sewage works [1]. Buff-banded Rail can often be seen along this creek. Brown Quail is sometimes also seen here but is more frequently found along the track through the nearby woodland [2]. White-throated Honeyeater and Pheasant Coucal are commonly seen in woodland around this track and Varied Lorikeet also occurs in this woodland when trees are flowering. Bush-hen has been seen here during the Wet Season.

The central track [3] allows views into the sewage works and though imperfect, may be an option if access into the sewage works is not available. Black Bittern may also be present during the Wet Season when there is tall grass immediately to the west of this

track [3] and Black-necked Stork is often seen in the shallow creek to the east.
Zitting Cisticola occurs in small numbers, particularly in the area around [4]. It is really
only easy to find this species in the Wet Season when it becomes more vocal and its
incessant 'tick-tick' call makes it conspicuous. Golden-headed Cisticola also occurs in
this locality, so any Cisticolas need to be observed carefully for accurate identification. Further along this track, Little Ringed Plover has been recorded around the
gravel piles [5] and, during the Wet Season, Yellow Wagtail can often be seen a little
beyond here at the edge of the mangroves. Flock Bronzewing sometimes occurs in the
grassy areas beyond [6] and it is presumably birds from here that are occasionally
seen in the sewage works during the Dry Season.
During the Wet Season a combination of high tide levels and heavy rainfall may
provide short-term habitat for shorebirds including Swinhoe's Snipe, Broad-billed
Sandpiper, Ruff and Long-toed Stint and it would be well worth checking the swamp
at these times.

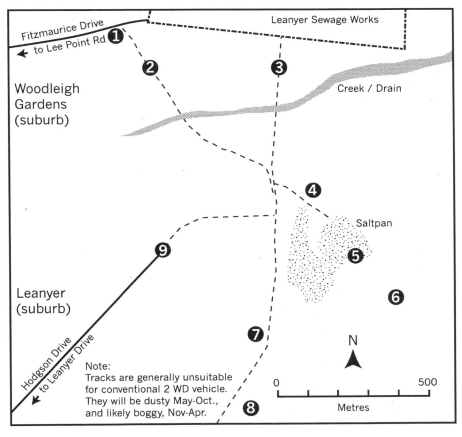

1.12 Holmes Jungle Nature Reserve

Holmes Jungle Nature Reserve is at the edge of Darwin's northern suburbs, about 20 km from the city. It comprises monsoon forest, swamp, open grassland and savanna woodland and holds a good range of birds in all these habitats. The swamp is usually best at the end of the Dry Season (Sep - Nov) if there is some shallow water with areas of mud and short grass. The Reserve is open to vehicles between 7 am and 7 pm.

Key Species
King Quail, Black Bittern, Little Bittern, Black Falcon, Spotted Harrier, Pacific Baza, Swinhoe's Snipe, White-browed Crake, Buff-banded Rail, Red-chested Button-quail, Red-backed Button-quail, Little Curlew, Oriental Pratincole, Rose-crowned Fruit-Dove, Emerald Dove, Rufous Owl, Eastern Grass Owl, Australian Owlet-nightjar, Rainbow Pitta, Zitting Cisticola

Other Species
Orange-footed Scrubfowl, Magpie Goose, Black Kite, Whistling Kite, Brown Falcon, Brolga, Australian Pratincole, Azure Kingfisher, Forest Kingfisher, Grey Whistler, Northern Fantail, Spangled Drongo, Red-backed Fairy-wren, Long-tailed Finch, Chestnut-breasted Mannikin

Finding Birds
Swamp / Grassland
Red-backed and Red-chested Button-quails may be seen in the small area of grassland [4] near the bottom of the hill. During the Wet Season, King Quail and Black Bittern may also be seen here.

Zitting Cisticola is normally easy to find by the fence [1] from November to March when it becomes active and calls frequently. To access the swamp [3] or the surrounding grassland, go through the gate near [1]. Red-backed and Red-chested Button-quails and King Quail may be seen in the long grass. It can be rough underfoot here, so take care. Black Falcon may be seen overhead, amongst the ever-present Black Kites, which congregate around the nearby rubbish tip. Spotted Harrier occasionally hunts over this area.

Little Curlew are often seen at the edge of the swamp, or in the grass a little further away. Australian Pratincole feed around the edges of the swamp. Oriental Pratincole, Black-tailed Godwit, Sharp-tailed Sandpiper, Wood Sandpiper and Marsh Sandpiper may be present when there is a good muddy edge. Brolga are usually found around the northern and eastern sides of the swamp, often in large numbers. White-browed Crake and Buff-banded Rail occur where there is a good mix of reeds and mud and Little Bittern has been recorded in the taller reeds during the Wet Season.

To access the stream and the grassland on its eastern side, park at the end of the track [P]. Rainbow Pitta, Azure Kingfisher, Emerald Dove and Grey Whistler may be seen in the monsoon forest at the edge of the stream. Brown and King Quails, Red-backed and Red-chested Button-quails and Swinhoe's Snipe occur in the grassy areas on the eastern side of the stream, either side of the track [5]. Eastern Grass Owl is rare but is sometimes seen at dusk hunting over the grasslands at [5] or beyond [2].

Monsoon forest

Park at [P1], which is near the start of the monsoon forest walking track. Walk through the picnic area, watching for Long-tailed Finch, Chestnut-breasted Mannikin and Red-backed Fairy-wren in the grass around the picnic area before getting to the forest boardwalk [6]. Rainbow Pitta, Rose-crowned Fruit-Dove, Rufous Owl and Australian Owlet-nightjar, along with the more common Grey Whistler, Spangled Drongo and Northern Fantail, may be seen almost anywhere along the forest tracks or boardwalks. Pacific Baza is most often seen at the monsoon forest edges. On the way out of the reserve, watch along the roadside [7] for Brown Quail, particularly late in the afternoon.

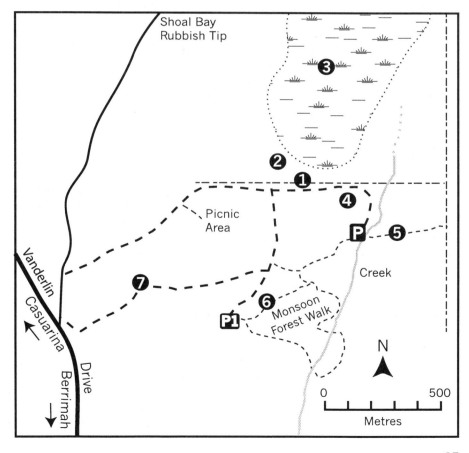

1.13 Knuckey Lagoons

Knuckey Lagoons Reserve is near Berrimah, approximately 15 km east of the city. About 150 species have been recorded here, with a large variety of waterfowl, herons and shorebirds, including a few national rarities. The reserve comprises three main lagoons surrounded by grassland. Birding is best from September to December, when waterbirds seek refuge from drying lagoons further inland and visitors and passage migrants such as Little Curlew, Oriental Plover and Oriental Pratincole may be present. Wet Season rains, usually from December, flood the lagoons and most of the birds move elsewhere.

Key Species
King Quail, Garganey, Glossy Ibis, Black-breasted Buzzard, Black Falcon, White-browed Crake, Red-backed Button-quail, Swinhoe's Snipe, Asian Dowitcher, Little Curlew, Ruff, Broad-billed Sandpiper, Pectoral Sandpiper, Long-toed Stint, Oriental Plover, Comb-crested Jacana, Bush Stone-curlew, Oriental Pratincole, Spotted Nightjar, Red-backed Kingfisher, Bar-breasted Honeyeater, Yellow Wagtail, Zitting Cisticola

Other Species
Australasian Grebe, Magpie Goose, Green Pygmy-goose, Wandering Whistling-Duck, Intermediate Egret, Black-necked Stork, Black-tailed Godwit, Wood Sandpiper, Marsh Sandpiper, Sharp-tailed Sandpiper, Black-fronted Dotterel, Australian Pratincole, Red-tailed Black-Cockatoo, Red-winged Parrot, Brush Cuckoo, Striated Pardalote, Blue-faced Honeyeater, Grey-crowned Babbler, Restless Flycatcher, Singing Bushlark

Finding Birds
All of the lagoons are productive for the more common waterbirds such as Wandering Whistling-Duck, Radjah Shelduck, Green Pygmy-goose, Magpie Goose, Black-necked Stork, Glossy Ibis, Intermediate Egret, Australian Pratincole and Comb-crested Jacana. The easiest (and often best) of the lagoons for birding is at the end of Fiddlers Lane [2]. Red-backed Kingfisher and Brush Cuckoo often perch on the power lines along Fiddlers Lane and Bush Stone-curlew is sometimes seen resting under trees in the mango orchard beside the lane. Little Curlew feed in the grassy area around the lagoon and Oriental Plover and Oriental Pratincole may be found on the mud at the water's edge or in the short grass a little further from the water. Wood Sandpiper can be numerous here and other shorebirds including Marsh and Sharp-tailed Sandpipers can usually be seen here and at Snipe Swamp [3].

Grey Butcherbird is often seen on the power line along Lagoon Road, at the Secrett Road end [4]. Singing Bushlark frequently perches on the fence beside the Snipe Swamp access track [1]. Bar-breasted Honeyeater can often be seen in the small clump of paperbark trees along to the left of the entry gate, and Zitting Cisticola may be found in the grassy areas between the gate and the swamp. Black Falcon and Black-breasted Buzzard occasionally hunt over the large open area surrounding this lagoon. King Quail and Red-backed Button-quail have been seen around the edges of the swamp and on the higher ground nearer the gate. Many shorebirds, including Oriental Pratincole, Swinhoe's Snipe, Little Curlew and Oriental Plover occur around Snipe Swamp from October to December, leaving when the water level rises. Ruff, Broad-billed Sandpiper and Long-toed Stint occur occasionally and Asian Dowitcher has also been recorded here. White-browed Crake is sometimes seen where the habitat is suitable. Garganey is recorded irregularly here and at the Stuart highway site.

The Stuart Highway site [3] has limited close viewing access, though there are often good numbers of birds there. Park on the track that runs beside the Stuart Highway or use the carpark in the adjacent plant nursery.

A number of areas beyond the reserve are also productive for birding. Striated Pardalote and Red-backed Kingfisher are often seen perched on the power lines by the Kentucky Turf Farm [5], and Yellow Wagtail is occasionally seen in the grass along the fence here. Bush Stone-curlew occurs in quite large numbers resting throughout the day amongst trees in front of the houses in Brandt Road [6].

The woodland around the Stevens and Campbell Road junction [7] is often good for Red-winged Parrot, Red-tailed Black-Cockatoo, Blue-faced Honeyeater and Grey-crowned Babbler. Spotted Nightjar is often seen along Campbell Road at night.

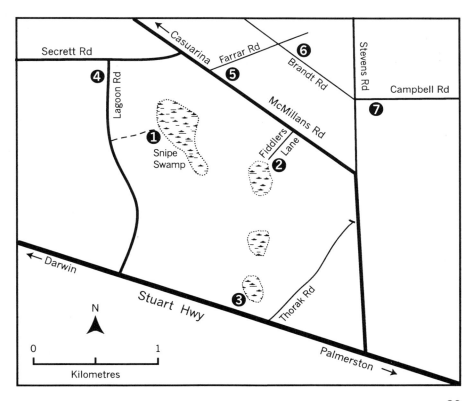

1.14 Palmerston Sewage Works

Palmerston is a rapidly developing city about 20 km from Darwin. The sewage works are at the end of Catalina Road, which runs off Elrundie Avenue. Over 200 species have been recorded in the relatively small area comprising mangroves, sewage ponds and surrounding woodland. This site is one of the most accessible and productive mangrove areas in the Top End and can be good even in the middle of the day. Note that the sewage works are not accessible to the public. It is however easy to view birds inside the sewage works through the fence.

Key Species
Brown Quail, Grey Goshawk, Chestnut Rail, Buff-banded Rail, White-browed Crake, Beach Stone-curlew, White-winged Black Tern, Diamond Dove, Little Bronze-Cuckoo, Spotted Nightjar, Little Kingfisher, Red-backed Kingfisher, Green-backed Gerygone, Large-billed Gerygone, Mangrove Gerygone, Red-headed Honeyeater, Broad-billed Flycatcher, Mangrove Robin, White-breasted Whistler, Mangrove Golden Whistler, Mangrove Grey Fantail, Yellow White-eye, Masked Finch, Yellow-rumped Mannikin

Other Species
Radjah Shelduck, White-bellied Sea-Eagle, Sharp-tailed Sandpiper, Terek Sandpiper, Pacific Golden Plover, Greater Sand Plover, Shining Flycatcher, Red-backed Fairy-wren, Striated Pardalote, Weebill, Helmeted Friarbird, Leaden Flycatcher, Crimson Finch, Long-tailed Finch, Chestnut-breasted Mannikin

Finding Birds
Red-backed Kingfisher regularly perches on power lines along Catalina Road during the Dry Season. Watch the woodland on the southern side of this road for Red-tailed Black-Cockatoo, Grey-crowned Babbler, Red-winged Parrot and Red-backed Fairy-wren. In some years Diamond Dove has also been seen here. Spotted Nightjar may be seen hunting around the road at night during the Dry Season.

Just short of the sewage works gate, there is a track [1] that follows the sewage works fence. Depending on the condition of the track it may be possible to drive as far as the corner [2]. Brown Quail, Crimson and Long-tailed Finches and Chestnut-breasted Mannikin are often seen in the woodland near the track and Masked Finch and Yellow-rumped Mannikin also occur here occasionally. White-bellied Sea-Eagle can often be seen perched atop one of the taller trees here. Grey Goshawk is frequently seen both in the woodland and the nearby mangroves.

White-browed Crake and Buff-banded Rail may be seen in the sewage works, wherever there is vegetation at the corners of the ponds. During the Wet Season, shorebirds such as Pacific Golden Plover, Sharp-tailed Sandpiper and Greater Sand Plover can usually be seen on the tracks between the ponds and White-winged Black Tern is often numerous. Freshwater Crocodiles are also regularly seen in the ponds.

At the far corner [3] of the sewage ponds fence there is a small but clear gap in the mangroves. Walk down into this gap a few metres and sit quietly. Patient observation may result in close views of Mangrove, Green-backed and Large-billed Gerygones, Red-headed Honeyeater, Yellow White-eye and Mangrove Grey Fantail. Mangrove Robin is often seen at the base of mangroves around the edge of the clearing. Little Kingfisher is sometimes also seen perched in the mangroves. Chestnut Rail, White-breasted and Mangrove Golden Whistlers and Broad-billed and Leaden Flycatchers have also been seen here.

Little Bronze-Cuckoo is generally easier to see in the mangroves nearer the sewage outflow pipe [4]. Helmeted Friarbird prefers the more open areas, generally at the edge of the mangroves [5].

Numerous shorebirds roost at high tide on the saltpans adjacent to the sewage works. To get good views of the saltpan, park near the large concrete pipe [6] and walk along the track [7]. Whimbrel is the commonest shorebird here but Lesser and Greater Sand Plovers, Pacific Golden Plover, Black-tailed Godwit and Terek Sandpiper also occur. Beach Stone-curlew is sometimes present, mainly late in the afternoon.

Grey-crowned Babbler, Red-tailed Black Cockatoo, Red-backed Fairy-wren and a good range of other woodland birds may be seen in the adjacent woodland [8].

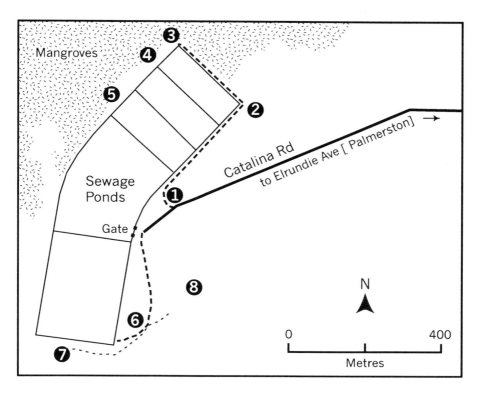

1.15 Elizabeth River & Channel Island

The Elizabeth River is about 6 km along the Channel Island road (Elrundie Ave) from Palmerston and provides easy access to mangroves with a good range of birds. Tide conditions are critical for Chestnut Rail and Great-billed Heron, which may be seen on the muddy banks when the tide is low.

Channel Island is about 17 km further on from the Elizabeth River bridge. It is at the end of the road and so cannot be done en route to other areas. The island holds a limited number of species but the mangroves and shore here provide one of the more reliable sites for Great-billed Heron. Varied Lorikeet and a range of honeyeater species may be seen in the woodland along the road between the Elizabeth River Bridge and Channel Island. Watch also for Black-tailed Treecreeper and Partridge Pigeon, which have occasionally been seen along this road.

Key Species

Great-billed Heron, Chestnut Rail, Partridge Pigeon, Emerald Dove, Rose-crowned Fruit-Dove, Varied Lorikeet, Rainbow Pitta, Black-tailed Treecreeper, Helmeted Friarbird, Silver-crowned Friarbird, Red-headed Honeyeater, Grey-crowned Babbler, Mangrove Robin, Yellow White-eye

Other Species

Eastern Reef Egret, Striated Heron, Osprey, Brahminy Kite, White-bellied Sea-Eagle, Rufous-banded Honeyeater, Varied Triller, Grey Butcherbird, Long-tailed Finch

Finding Birds

From the Elizabeth River bridge, or the adjacent boat ramp [3], scan the muddy banks of the river for Great-billed Heron. Chestnut Rail can sometimes be seen where there are views further into the mangroves. It may also be seen in the mangroves around the electricity pylon [1], where there is a small tidal creek. Mangrove Robin, Red-headed Honeyeater and Helmeted Friarbird may be seen in mangroves along the track [2], or around the pylon.

Varied Lorikeet, Silver-crowned Friarbird, Grey-crowned Babbler, Long-tailed Finch and Grey Butcherbird can often be seen in the roadside woodland between Elizabeth River and Channel Island. Black-tailed Treecreeper and Partridge Pigeon have also been seen along here though both are very uncommon in this area.

From the bridge connecting to Channel Island, look along the river banks for Great-billed Heron. Also check the electricity pylons [9] in the water as the herons occasionally perch on them, or their concrete bases, at high tide. Follow the road onto the island and immediately turn left. Rainbow Pitta, Emerald Dove and Rose-crowned Fruit-Dove are sometimes seen in the woodland/mangrove ecotone by the road junction [8]. At the end of the bitumen road there is a rather tangled network of dirt tracks. It is probably best to park in the boat ramp carpark [4] and walk to the point [5], where there are good views of the estuary. Great-billed Heron often feeds on the large expanses of mud [6] at low tide. At high tide it often roosts, sometimes well hidden, in the large mangrove trees along the river bank [7].

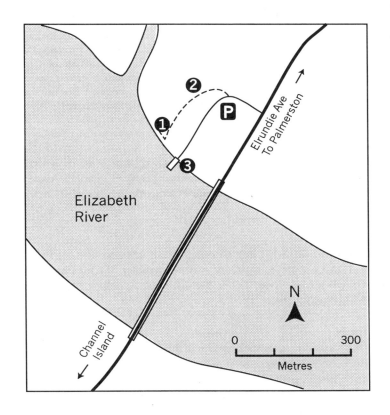

Elizabeth
River

② ①

P

③

Elrundie Ave
To Palmerston

Channel
Island

N

0   300
Metres

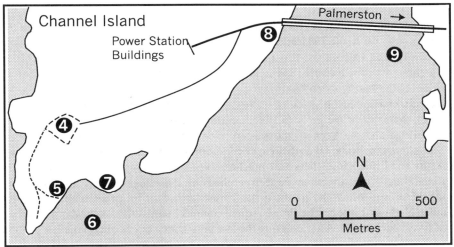

Channel Island

Power Station
Buildings

⑧

Palmerston →

⑨

④

⑤ ⑦

⑥

N

0   500
Metres

1.16 Howard Springs Nature Reserve

Howard Springs Nature Reserve is about 5 km from the Stuart Highway, a little over 30 km southeast of Darwin and about 12 km east of Palmerston. It is open from 8 am to 8 pm. The spring in the Reserve provides water that sustains an excellent area of monsoon forest, through which there is an easy loop walk of about 1.8 km length. A little over 100 species have been recorded in the Reserve's monsoon forest and surrounding woodland. Howard Springs Nature Reserve is consistently the easiest area in Darwin, and perhaps the entire Top End, to find Rainbow Pitta.

Key Species
Black Bittern, Pacific Baza, Grey Goshawk, Emerald Dove, Rose-crowned Fruit-Dove, Rufous Owl, Barking Owl, Large-tailed Nightjar, Tawny Frogmouth, Azure Kingfisher, Little Kingfisher, Rainbow Pitta, Large-billed Gerygone, Bar-breasted Honeyeater, Little Shrike-thrush

Other Species
Orange-footed Scrubfowl, Little Corella, Brush Cuckoo, Rufous-banded, White-throated and Dusky Honeyeaters, Grey-crowned Babbler, Shining Flycatcher, Spangled Drongo, Grey Whistler, Varied Triller, Yellow Oriole, Crimson, Double-barred, Long-tailed and Masked Finches

Finding Birds
Little Shrike-thrush is sometimes present in the dense growth immediately beside the car park [P1]. Emerald Dove, and sometimes Rainbow Pitta, may be seen on the lawns around the children's pool [5], in the early morning or late afternoon. Yellow Oriole, Brush Cuckoo, Rose-crowned Fruit-Dove and Varied Triller are frequently seen in the trees in the picnic area [6] and Rainbow Pitta sometimes ventures out of the forest onto the lawn here.

At the bridge [1], scan the far edges of the pool for Little Kingfisher. Also check the stream sides for Azure and Little Kingfishers, Shining Flycatcher and Large-billed and Green-backed Gerygones. It is generally best to walk clockwise around the track, watching out for Rainbow Pitta right from the start. If one has not been seen before reaching the first wooden footbridge [2], wait at the bridge for a while and listen for the call, or for the soft rustle of leaves as the Pitta hops about on the ground. Also check the sides of the gully (a stream during the Wet Season) on both sides of the footbridge and the trees on the western side, where the Pitta sometimes perches. Throughout the walk, take advantage of any opportunity to gain access to the stream. Azure and Little Kingfishers are often heard as they fly along the stream and, as with Black Bittern, can sometimes be seen perched on a fallen log or on a branch overhanging the stream. Watch the ground in the denser parts of the forest for a roosting Large-tailed Nightjar. Little Shrike-thrush is quite common, particularly in the denser pockets of forest such as [3]. After crossing the stream the area opens out on the left [4] and it is in this area that it is worth watching out for finches and honeyeaters. Crimson, Double-barred, Long-tailed and Masked Finches and Chestnut-breasted Mannikins can be seen here. White-throated and Rufous-banded Honeyeaters are usually quite common and Dusky and Bar-breasted Honeyeaters may also be seen. Rather than to continue walking along the loop, it is often better to return the same way, giving a second chance for species missed on the first part of the walk. It may be that near constant looking at the forest floor for Rainbow Pitta has meant missed opportunities for birds such as Pacific Baza, Grey Goshawk, Barking and Rufous Owls and Tawny Frogmouth, which may be perched quietly in the trees.

An alternative site for Rainbow Pitta is the area immediately around the spring [7]. Rose-crowned Fruit-Dove can often be seen here also. A range of the commoner woodland birds may be seen in the woodland to the east of the reserve [8].

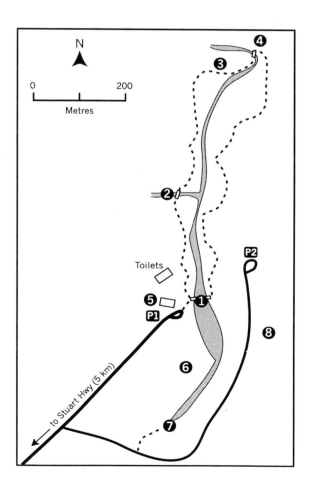

1.17 McMinns Lagoon

McMinns Lagoon lies in the area bounded by the Stuart and Arnhem Highways, about 35 km from Darwin. A little over 100 species have been recorded here. It is generally best for waterbirds around October, though earlier can still be very productive depending on water levels in other areas. From October, until the Wet Season rains fill the lagoon, the edges may be shallow and muddy, providing habitat for Swinhoe's Snipe and Oriental Pratincole.

Key Species
Garganey, Green Pygmy-goose, Black Bittern, Black-breasted Buzzard, Black Falcon, Swinhoe's Snipe, Diamond Dove, Cockatiel, Varied Lorikeet, Northern Rosella, Pheasant Coucal, Black-eared Cuckoo, Little Bronze-Cuckoo, Brush Cuckoo, Azure Kingfisher, Silver-crowned Friarbird, Tawny Grassbird

Other Species
Magpie Goose, Wandering Whistling-Duck, Pied Heron, Comb-crested Jacana, Black-winged Stilt, Little Curlew, Marsh Sandpiper, Wood Sandpiper, Sharp-tailed Sandpiper, Horsfield's Bronze-Cuckoo, White-throated Honeyeater, Rufous Whistler, Restless Flycatcher, White-winged Triller, Olive-backed Oriole, Grey Butcherbird

Finding Birds
From the carpark [P1] follow the well-defined track that leads to the lagoon. A range of honeyeater species along with Olive-backed Oriole and Varied Lorikeet may be seen in the trees along the track. Horsfield's Bronze-Cuckoo, Brush Cuckoo and, mainly during drier years, Black-eared Cuckoo, can also be found here. There is a poorly marked track [3] going east through the woodland and this can also be worthwhile. Little Bronze-Cuckoo, Pheasant Coucal, Red-winged Parrot, Northern Rosella and Silver-crowned Friarbird are likely along here and there is the possibility of Diamond Dove and Cockatiel during drier years. Watch the sky above and around the lagoon for raptors, including Black-breasted Buzzard and Black Falcon.

A small shelter [1] at the end of the track provides a view of the lagoon, so that waterbirds can be seen with a spotting scope from the shade. Common waterbirds include Green Pygmy-goose, Magpie Goose, Wandering Whistling-Duck, Pied Heron and Comb-crested Jacana. Garganey occurs irregularly. Black Bittern is frequently seen in the reeds at the edge of the lagoon [2] and Tawny Grassbird also occurs in this habitat. Azure Kingfisher is often seen in the vegetation by the shelter. It is possible to walk round to the back of the lagoon though this is easier from the alternative parking place [P2]. This will allow closer views of birds in that part of the swamp and will also increase the chance of seeing Swinhoe's Snipe. A range of woodland birds including honeyeaters, Restless Flycatcher, Rufous Whistler and Grey Butcherbird may be seen along the short track between carpark [P2] and the shelter [1].

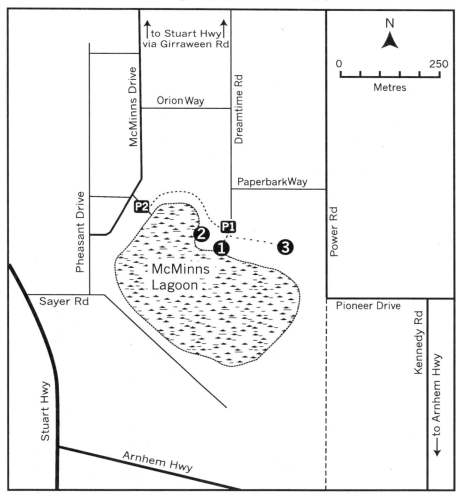

1.18 Darwin River Dam

Darwin River Dam is a little over 60 km from Darwin. It is one of the few waterbodies near Darwin where Great Crested Grebe and Great Cormorant occur regularly, although the range of waterbirds tends to be low. It is generally a reliable site for Northern Rosella.

Key Species
Great Crested Grebe, Great Cormorant, Northern Rosella, White-browed Robin, Cicada-bird, Green-backed Gerygone

Other birds
Torresian Imperial-Pigeon, Red-tailed Black-Cockatoo, Great Bowerbird, White-throated Honeyeater, Little Friarbird, Shining Flycatcher, Crimson Finch, Masked Finch

Finding Birds
White-browed Robin and Cicadabird have been seen in the dense riverine vegetation near the entrance gate [1]. Suitable habitat for these species can also be found at the rear of the large parkland area [2] or at the end of the short track beside the dam wall [6]. To access the dam itself, park in the designated area [P]. Northern Rosella is commonly seen in the trees around the picnic area [3]. Check here also for Green-backed Gerygone and Great Bowerbird. Great-crested Grebe is usually some distance out on the dam. Great Cormorant is often seen on the rocky dam edges beside the observation tower [4] or further along near the boat ramp [5].

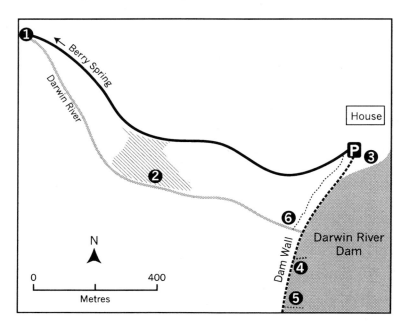

2 Fogg Dam Region

The Fogg Dam Region covers the area along the Arnhem Highway, from Fogg Dam to the Mary River, which is approximately 110 km from Darwin. It is characterised by subcoastal wetlands, including the Mary and Adelaide River floodplains and some large lagoons and billabongs. The region is easily accessed as a day trip from Darwin, or it can be visited on the way to or from Kakadu National Park. An alternative is to stay at Mary River Park [site 2.06], an eco-tourism resort, and use this as a base from which to explore the region.

About 230 bird species have been reported from this region. Apart from the extensive wetlands, there are patches of monsoon forest containing birds such as Rainbow Pitta and Emerald Dove. Seasonally, Oriental Plover and Little Curlew can be seen on the open grasslands in this region and in some years Letter-winged Kite may also be seen hunting over the floodplains and grasslands.

The Marrakai Road, a particularly productive area for birding, runs between the Arnhem and Stuart Highways.

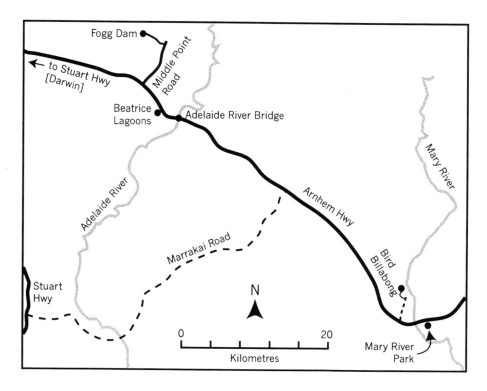

2.01 Fogg Dam

Fogg Dam is about 52 km east of Darwin, the turn-off 32 km along the Arnhem Highway from the Stuart Highway. It comprises a large wetland area and extensive monsoon forest. There are observation platforms along the dam wall and a large observation tower at its end. The dam wall gives excellent views of the wetlands and is one of the easiest places in the Top End to see White-browed Crake. There are well marked walking tracks in the adjacent monsoon forest. At night, spotlighting along the dam wall can be particularly productive.

Key Species
Little Bittern, Pacific Baza, Grey Goshawk, White-browed Crake, Baillon's Crake, Buff-banded Rail, Red-backed Button-quail, Channel-billed Cuckoo, Oriental Cuckoo, Brush Cuckoo, Little Bronze-Cuckoo, Barking Owl, Spotted Nightjar, Large-tailed Nightjar, Emerald Dove, Rose-crowned Fruit-Dove, Little Kingfisher, Azure Kingfisher, Rainbow Pitta, Bar-breasted Honeyeater, Little Shrike-thrush, Broad-billed Flycatcher, Olive-backed Oriole, Tawny Grassbird, Barn Swallow

Other Species
Magpie Goose, Green Pygmy-goose, Radjah Shelduck, Pied Heron, Little Egret, Cattle Egret, Royal Spoonbill, Glossy Ibis, Black-necked Stork, Swamp Harrier, Purple Swamphen, Comb-crested Jacana, Bush Stone-curlew, Australian Pratincole, Whiskered Tern, Tawny Frogmouth, Dusky Honeyeater, White-gaped Honeyeater, Restless Flycatcher, Shining Flycatcher, Lemon-bellied Flycatcher, Spangled Drongo, Varied Triller, Yellow Oriole, Crimson Finch, Clamorous Reed-Warbler

Finding Birds
Bush Stone-curlew, Green-backed Gerygone and Lemon-bellied, Leaden and Broad-billed Flycatchers may be seen around the main carpark [P1]. Raptors including Pacific Baza and Grey Goshawk can also be seen. Barking Owl starts to call around dusk and is frequently seen near the carpark. Spotted Nightjar hawks for insects around the carpark and along the dam wall from dusk. Large-tailed Nightjar may also be seen here, particularly during the Wet Season.

The dam wall allows excellent views of the wetland on one side and the grassy plain on the other. Bar-breasted Honeyeater and sometimes Oriental Cuckoo can be seen in the trees or bushes along the start of the dam wall. When there is surface water, check the open area [4] near the forest edge for crakes. White-browed Crake, which is generally wary, sometimes comes out in full view of the first viewing platform [5]. Watch the base of the surrounding reeds for Baillon's Crake, which is also a possibility though less common. Little Bittern and Oriental Reed-Warbler, both rarities, have been seen in the reeds. Clamorous Reed-Warbler and Tawny Grassbird are common and may be seen on either side of the dam wall. Magpie Goose, Brolga and Black-necked Stork as well as many other waterbirds may be seen scattered throughout, mainly on the north-ern side of the wall. The large observation deck [6] provides views across the extensive grassland. When there are areas of shallow water with muddy edges watch for Buff-banded Rail. Crocodiles are also often seen from this deck.

There are two monsoon forest walks. However, if there is only time for one, the southern walk, signposted "Woodland to Waterlilies Walk" is usually better for birdwatching and is somewhat shorter. At the start of this walk a small creek is crossed. Watch towards the dam edge here for Little and Azure Kingfishers. Rainbow Pitta, Brush Cuckoo, Bar-breasted Honeyeater and Leaden and Broad-billed Flycatchers are also often seen along this part of this walk, together with Little Shrike-thrush, Yellow Oriole, Spangled Drongo and Lemon-bellied Flycatcher. The track changes to a board-walk further along and for much of the year the ground in this section is under water. Little and Azure Kingfishers are sometimes found perched on branches above the water here. Rose-crowned Fruit-Dove is often in the trees at the edge of the forest, where the track passes through an open area before reaching the small observation deck [2] at the wetland edge. White-browed Crake can be found on or around the waterlilies near the observation deck.

The "Monsoon Forest Walk" passes first through woodland and Olive-backed Oriole is often seen here. Rose-crowned Fruit-Dove and Rainbow Pitta have been seen throughout the monsoon forest, though the Pitta is more frequently found within the loop part of the walk. A range of waterbirds may be seen on the wetland from [3].

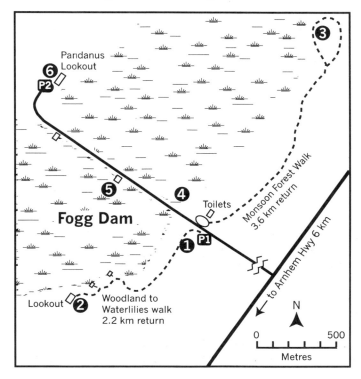

2.02 Middle Point Road

Middle Point Road runs between the Arnhem Highway and Fogg Dam. It passes through open grassland with sparse trees, to fairly dense woodland. About 85 bird species have been recorded here, including Letter-winged Kite, which may occur when there is an irruption from its inland breeding areas. It is an excellent site for night time spotlighting.

Key Species
Black-breasted Buzzard, Spotted Harrier, Letter-winged Kite, Black Falcon, Oriental Plover, Australian Pratincole, Little Curlew, Oriental Plover, Brush Cuckoo, Oriental Cuckoo, Barking Owl, Barn Owl, Masked Owl, Spotted Nightjar, Red-backed King-fisher, Black-faced Woodswallow, Rufous Songlark, Singing Bushlark, Barn Swallow

Other Species
Cattle Egret, Wedge-tailed Eagle, Black-shouldered Kite, Nankeen Kestrel, Brown Falcon, Australian Hobby, Blue-winged Kookaburra, Forest Kingfisher, Richard's Pipit, Long-tailed Finch, Masked Finch

Finding Birds
Black-faced Woodswallow can be seen near the Fogg Dam - Arnhem Highway junction, which is about the closest to Darwin that this species is normally found. Brown and Rufous Songlarks, Singing Bushlark, Little Curlew, Oriental Plover and Australian Pratincole are most frequently seen at the Arnhem highway end of the road [1-3], particularly on the western side. Red-backed Kingfisher may be seen perched on power lines here throughout the year. Oriental Cuckoo also perches on the power line where the vegetation is denser [4] and on the power line along the Harrison Dam road [6]. In some years Letter-winged Kite may be seen perched in trees in the paddocks [3] or, particularly around dawn or dusk, on any of the tall communication masts [2] on the opposite side of the road. After dusk, Barking Owl and Tawny Frogmouth often perch on power lines north of the dam turn-off [5]. Barn Owl and Spotted Nightjar are mainly seen at the southern end of the road [3 - 4]. Masked Owl has been recorded in this area but is very rare.

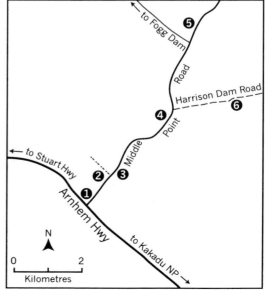

2.03 Beatrice Lagoons

Beatrice Lagoons are along the Arnhem Highway, approximately 28 km from the Stuart Highway. These lagoons attract large numbers of waterbirds as other wetlands dry up. About 100 species have been recorded at the lagoons and in the immediate vicinity.

Key Species

Garganey, Plumed Whistling-Duck, Green Pygmy-goose, White-browed Crake, Spotless Crake, Buff-banded Rail, Comb-crested Jacana, Little Curlew, Oriental Pratincole, Red-necked Avocet

Other Species

Magpie Goose, Wandering Whistling-Duck, Intermediate Egret, Nankeen Night Heron, Brolga, Brush Cuckoo, Forest Kingfisher

Finding Birds

The most productive areas vary from year to year depending on the amount and distribution of water in the lagoons. Both Wandering and Plumed Whistling-Ducks are usually present in the southern section of the lagoon [1], along with Radjah Shelduck and Green Pygmy-goose. Garganey has also been recorded here. Herons, shorebirds, White-browed Crake, Buff-banded Rail and Comb-crested Jacana are more likely in the central part of the lagoons [3]. Spotless Crake has been seen on the muddy edges of the small pond on the eastern side of the road [2] but ideal conditions there occur infrequently. Little Curlew and Oriental Pratincole can be seen around the southern end of the lagoons [1] when the edges are muddy and the surrounding grass is short.

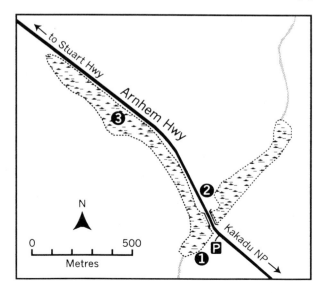

2.04 Adelaide River Bridge

The Adelaide River bridge site is located about 7 km past the Fogg Dam turn-off along the Arnhem Highway. Do not confuse it with the town of Adelaide River, which is on the Stuart Highway. This site is probably the easiest and most reliable site in the Top End for Mangrove Golden Whistler.

Key Species

Emu, Great-billed Heron, Letter-winged Kite, Oriental Cuckoo, Little Bronze-Cuckoo, Rainbow Pitta, Green-backed Gerygone, Red-headed Honeyeater, Mangrove Golden Whistler, Broad-billed Flycatcher, Rufous Fantail, Yellow White-eye

Other Species

Brolga, Brush Cuckoo, Forest Kingfisher, Shining Flycatcher, Restless Flycatcher, Lemon-bellied Flycatcher, Crimson Finch, Long-tailed Finch

Finding Birds

Park at the Adelaide River Queen carpark. Mangrove Golden Whistler can be seen almost anywhere along the edge of the river but the best place to start looking is between the bridge and the tour centre buildings [2]. If the gate to the enclosure [3] is open, look there for the Whistler, Rainbow Pitta, Green-backed Gerygone and Rufous Fantail. Oriental Cuckoo and Little Bronze-Cuckoo may be seen along track [4] or, along with Mangrove Golden Whistler, in the scrubby vegetation by the river at the end of the track [5]. Restless, Shining and Broad-billed Flycatchers, Red-headed Honeyeater and Yellow White-eye may also be seen here or along the river's edge.

Black-breasted Buzzard and sometimes Letter-winged Kite may be seen flying over the nearby floodplain [6]. Brolga is often on the floodplain on either side of the road and there have been a few records of Emu. Great-billed Heron can sometimes be seen on the river banks at low tide.

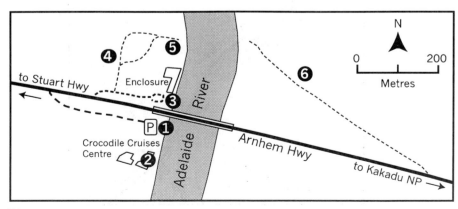

2.05 Marrakai Track

The unsealed Marrakai Track runs between the Arnhem Highway, about 18 km southeast of the Adelaide River bridge, and the Stuart Highway, about 40 km south of its junction with the Arnhem Highway. A 4WD vehicle is required where the track fords the Adelaide River [4]. However, the Arnhem Highway end of the track [1 - 4] provides the track's best birding and, provided there has been no significant incidental damage to the road, can be done in a conventional vehicle.

Key Species
Chestnut-backed Button-quail, Little Eagle, Black-breasted Buzzard, Spotted Harrier, Black Falcon, Grey Falcon, Diamond Dove, Horsfield's Bronze-Cuckoo, Black-tailed Treecreeper, Varied Sittella, Singing Honeyeater, White-browed Robin, Jacky Winter, White-throated Gerygone

Other Species
Common Bronzewing, Northern Rosella, Red-backed and Azure Kingfishers, Rufous Whistler, Grey-crowned Babbler, Rufous-throated Honeyeater, Zitting Cisticola

Finding Birds
Stop at intervals and check the woodland, as almost any section can be very productive. Chestnut-backed Button-quail, Black-tailed Treecreeper and Banded Honeyeater may be found in the woodland around [2; approx 4.5 km from 1]. A number of raptor species have been seen hunting over the floodplain [3; approx 14 km from 1] or perched in trees at its edge. Diamond Dove, Horsfield's Bronze-Cuckoo, Singing and Rufous-throated Honeyeaters also occur here. White-browed Robin and Shining and Restless Flycatchers can be seen at the crossing [4: approx 30.5 km from 1].

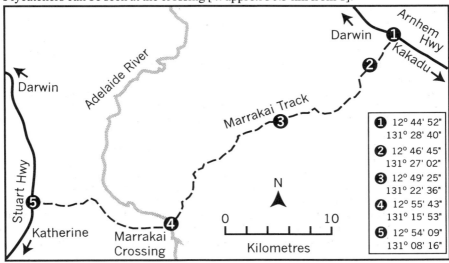

2.06 Mary River Park

Mary River Park is an eco-tourism resort on the Arnhem Highway, 110 km from Darwin. Nearly 200 species have been recorded in and around the area. The resort is inexpensive and has cabin accommodation, camping and Youth Hostel dormitory accommodation, bar and restaurant, making a birdwatching stay comfortable. River cruises are available, offering a good opportunity to see a range of birds, including Great-billed Heron and Black Bittern as well as Estuarine and Freshwater Crocodiles. Guided birding tours of the park and other areas are also available. See *Further Information* (p.160) for Mary River Park contact details.

The nearby Bird Billabong Reserve holds large numbers of waterbirds during the Dry Season. It is north of the Arnhem Highway, about 2 km west of Mary River Park, or 41 km east of the Adelaide River. The road to the carpark (3 km) is unsealed but well maintained. The walk to the billabong observation deck is 1.4 km from the carpark.

Key Species
Great-billed Heron, Black Bittern, Grey Goshawk, Pacific Baza, Square-tailed Kite, Australian Bustard, Red-necked Avocet, Oriental Cuckoo, Rufous Owl, Large-tailed Nightjar, Partridge Pigeon, Rose-crowned Fruit-Dove, Varied Lorikeet, Cicadabird, White-browed Robin, Zitting Cisticola

Other Species
White-bellied Sea-Eagle, Swamp Harrier, Red-tailed Black-Cockatoo, Dollarbird, Little Shrike-thrush, Bar-breasted Honeyeater, Black-tailed Treecreeper, Broad-billed Flycatcher, White-throated Gerygone, Green-backed Gerygone, Masked Finch, Chestnut-breasted Mannikin

Finding Birds
There are marked walks starting at the end of the Caravan Park [1, 2].
Bamboo walk [1]
White-browed Robin is often present near the start of the bamboo walk. Watch quietly and patiently for it in the lower half of the canopy. Little Shrike-thrush, Shining Flycatcher, Northern Fantail, Green-backed and White-throated Gerygones and Spangled Drongo also occur here. Large-tailed Nightjar is often seen in the densest part of the Bamboo walk. Rufous Owl can sometimes also be seen in here, or swooping over the billabong. As the track passes from the bamboo area to riverine forest watch in the flowering *Eucalyptus* and *Melaleuca* trees for Varied Lorikeet and for a range of honeyeaters. White-throated, Dusky, Blue-faced, Bar-breasted, Banded and Rufous-throated Honeyeaters, along with the more common Rufous-banded, White-gaped and Brown Honeyeaters and Silver-crowned Friarbird are found here. Pacific Baza, Brown Goshawk and Red-winged Parrot can also be seen in this section.
Wallaby Track [2]
Rainbow Pitta is an occasional visitor to this section, which includes remnants of monsoon forest. Barking Owl can be seen roosting during daylight here and is also a regular visitor to the lights of the Caravan Park, feasting on the insects at night.
Northern Rosella occurs here from about May onwards. Red-backed Fairy-wren is frequently found at the end of the walk amongst the spear grass. Partridge Pigeon can often be seen around the edges of the path from the gate entrance to the car park.

Boat Cruise

Great-billed Heron is regularly seen from April to October, after the Wet Season floods have receded. Black Bittern nests on the river and is seen all year round. Pacific Baza and the white morph of the Grey Goshawk are frequently seen along the river. Forest, Sacred and Azure Kingfishers are commonly present along the banks of the river while Little Kingfisher is sometimes also seen. Oriental Cuckoo may be seen during the Wet Season. Flocks of Varied Lorikeet feast on the flowering *Melaleuca* trees in June and July.

Bird Billabong

Masked Finch can be found in woodland around the carpark [P]. Varied Lorikeet, Bar-breasted and other honeyeater species feed in the flowering paperbark trees along and near the boardwalk [5 - 6]. Black-tailed Treecreeper is sometimes present around the carpark, but is easier to find in the woodland [7 - 8] further toward the billabong, where Brush Cuckoo, Boobook Owl, Rufous Whistler and other woodland birds occur. Where there is dense grass Chestnut-breasted Mannikin and Masked and Crimson Finches may be seen. The covered viewing area [9] gives expansive views of the billabong, where a range of waterbirds can be seen during the Dry Season. A number of shorebirds including Red-necked Avocet have also been recorded here.

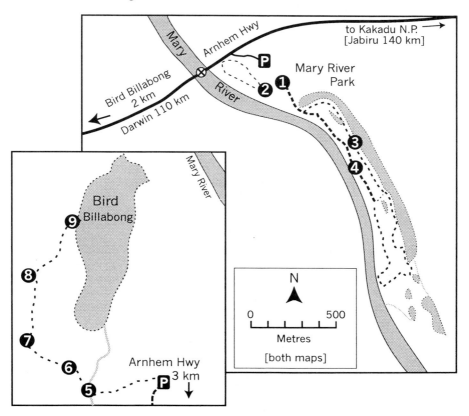

3 Kakadu National Park

Kakadu National Park is approximately 250 km east of Darwin via the Arnhem Highway or 60 km north of Pine Creek via the Kakadu Highway. The Park contains a diverse range of habitats including open woodlands, monsoon forests, sandstone escarpment, floodplains and billabongs.

Approximately 300 species of birds have been recorded within the Park, including Chestnut-quilled Rock-Pigeon, Banded Fruit-Dove and White-throated Grasswren, which are only found within or near Kakadu National Park.

The Bowali Visitors Centre at Park Headquarters, near the junction of the Kakadu and Arnhem Highways, and Warrdjan Aboriginal Cultural Centre at Cooinda, are both excellent sources of information on all aspects of the Park. Due to its size it is not feasible to cover Kakadu National Park in one day. There are several campgrounds within the Park and cabin and motel accommodation can be found at South Alligator, Jabiru, Cooinda and Mary River Roadhouse. There is a supermarket at Jabiru, where supplies can be purchased on weekdays or Saturday mornings.

The Park entry fee is valid for 7 days. A guidebook that includes reasonable maps and other information is provided on payment of the entry fee.

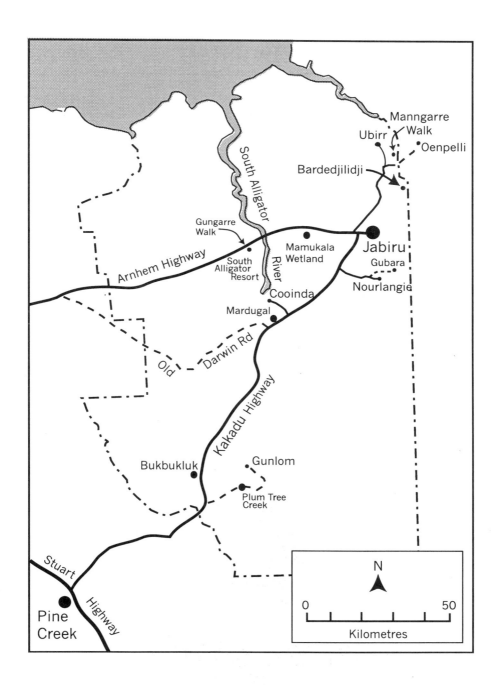

3.01 South Alligator River

The South Alligator River crosses the Arnhem Highway about 80 km east of the Kakadu National Park boundary and 40 km west of Jabiru. The river is lined with mangroves, beyond which is an extensive floodplain. Yellow Chat has been recorded breeding on the floodplain here. The Gungarre Walk passes through monsoon forest near the South Alligator Resort. Camping and cabin accommodation are available at the South Alligator Resort (Frontier Kakadu Village).

Key Species
King Quail, Spotted Harrier, Letter-winged Kite, White-browed Crake, Little Curlew, Oriental Cuckoo, Rufous Owl, Barking Owl, Eastern Grass Owl, Large-tailed Nightjar, Rainbow Pitta, Yellow Chat, Little Shrike-thrush, Tawny Grassbird, Zitting Cisticola

Other Species
Royal Spoonbill, Black-necked Stork, Swamp Harrier, Black-shouldered Kite, Red-kneed Dotterel, Little Corella, Brush Cuckoo, Barn Owl, Tawny Frogmouth, Australian Owlet-nightjar, White-bellied Cuckoo-shrike, Yellow Oriole, Figbird, Golden-headed Cisticola, Chestnut-breasted Mannikin

Finding birds
During the Wet Season, drive slowly along the road through the floodplain [5 - 6], with the vehicle windows down and listen for the distinctive call of Zitting Cisticola. Once heard, this species can be located easily. Tawny Grassbird and Crimson Finch can also be seen in the low floodplain vegetation at the edges of the roadside. White-browed Crake can sometimes be seen in gaps in the vegetation and King Quail and Red-backed Button-quail have also been recorded here. Swamp Harrier and Black-shouldered Kite are seen quite commonly over the floodplain.

During the Dry Season, Brolga and Australian Pratincole are often present on the dry grassland and Black-breasted Buzzard and Spotted Harrier are seen here occasionally. Letter-winged Kite has been recorded here and there are also records of breeding colonies of this species in the area during periods of irruption.

Yellow Chat has been recorded on the floodplain [5] and the species may still be in the general area, so it is worthwhile watching out for it. Eastern Grass Owl has been seen hunting over the floodplain. Barn Owl also occurs so take care with identification. Great-billed Heron is quite common along the river and is sometimes seen on the river banks here at low tide. Although there is little suitable access to the river on the north side of the highway, it is best to park in the main car park [P] on that side and walk across to [6], which provides good views of the river bank.

The Gungarre Walk [2] starts near the South Alligator Resort [1]. Much of it passes through monsoon forest in which Rufous Owl, Oriental Cuckoo, Rainbow Pitta and Little Shrike-thrush occur. At [4] the track passes alongside an extensive swamp. Magpie Goose, Wandering and Plumed Whistling-Duck, Glossy Ibis, Black-necked Stork and many other waterbirds can be seen here when water levels are suitable. Barking Owl, Rufous Owl and Large-tailed Nightjar, along with the more common Tawny Frogmouth and Australian Owlet-nightjar, may be found by spotlighting along the road or around the Gungarre Walk at night.

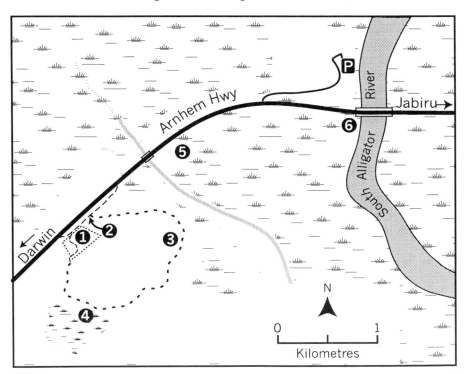

3.02 Mamukala Wetlands

Mamukala is about 1 km south of the Arnhem Highway. The turn-off is well signposted, 31 km west of Jabiru or 86 km east of the Arnhem Highway entrance to Kakadu National Park. As many wetlands in the region dry up Mamukala becomes more important as a refuge for water-birds. By October there is usually excellent shorebird habitat, which is used by migrants such as Little Curlew and Oriental Pratincole.

Key Species
Black-breasted Buzzard, Square-tailed Kite, Little Curlew, Oriental Plover, Oriental Pratincole, Barking Owl, Black-eared Cuckoo, White-browed Robin

Other Species
Magpie Goose, Green Pygmy-goose, Black-necked Stork, Glossy Ibis, Brown Falcon, Little Corella, Red-winged Parrot, Lemon-bellied Flycatcher, Leaden Flycatcher, Restless Flycatcher, Varied Triller, Masked Finch, Long-tailed Finch

Finding Birds
There are two bird hides at Mamukala. The first [1] is about 200 m from the carpark. It is best to visit this hide in the early morning before it becomes busy, particularly from May to September. Bird activity in the woodland by the track is much greater in the early morning. White-browed Robin, Lemon-bellied Flycatcher, Restless Fly-catcher and Leaden Flycatcher occur in the denser vegetation just before the hide.

The walk to the other bird hide [2] is also well worthwhile. Masked and Long-tailed Finches are common along this walk, as are Red-winged Parrot, Varied Triller and Banded Honeyeater. Barking Owl, White-throated Gerygone and Black-eared Cuckoo are also sometimes seen here.

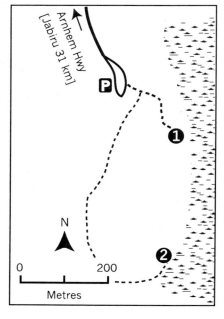

Glossy Ibis, Brolga, Black-necked Stork, Green Pygmy-goose, Plumed and Wandering Whistling-Ducks and Comb-crested Jacana are common on the wetland and build up in numbers as the dry weather progresses. From October to December, Mamukala Wetland dries up, leaving extensive areas of mud around the edges. Little Curlew, Oriental Plover and Oriental Pratincole can sometimes then be seen in front of the hides. Black Falcon, Black-breasted Buzzard and Square-tailed Kite are quite frequently recorded in this area.

3.03 Manngarre Walk

The Manngarre Walk starts opposite the Border Store, which is 37 km north of the Arnhem Highway along the Ubirr (Oenpelli) Road. It is a 1.6 km circular walk passing through a small monsoon rainforest adjacent to the East Alligator River. Be aware that Estuarine Crocodiles are common in the river here. There is camping at Merl, 2 km west of the start of the walk.

Key Species
Grey Goshawk, Pacific Baza, Rose-crowned Fruit-Dove, Little Bronze-Cuckoo, Oriental Cuckoo, Rufous Owl, Barking Owl, Rainbow Pitta, Little Shrike-thrush

Other Species
Orange-footed Scrubfowl, Torresian Imperial-Pigeon, Common Koel, Dollarbird, White-gaped Honeyeater, Varied Triller, Figbird, Yellow Oriole

Finding Birds
Rainbow Pitta is often seen in the first 400 m of the walk. It is also frequently found near [3] and in the area around the raised walkway [4 - 5]. Little Shrike-thrush may be seen anywhere along the walk but especially where the vegetation is dense. Listen for its distinctive call. Listen also for Rose-crowned Fruit-Dove calling or feeding (i.e. dropping fruit) within the forest. Check fruiting figs carefully for this species, as it often sits quietly and unobtrusively. Little Bronze-Cuckoo, Oriental Cuckoo, Common Koel and Yellow Oriole are also frequently seen along this walk. The more common species include White-gaped Honeyeater, Varied Triller, Figbird and Dollarbird. Watch for Black Bittern and Great-billed Heron where the walk comes close to the East Alligator River [1 - 2]. In the early morning these species feed along the river, then tend to retreat into the smaller, more sheltered creeks later in the day as it becomes hotter. Great-billed Heron has also been seen from the boat ramp [7] in the early morning at low tide. Grey Goshawk is often reported from the edge of the forest, where the path crosses an extensive open rocky area [6].

Spotlighting at night along the loop track and the carpark around the Border Store can be worthwhile, as both Rufous Owl and Barking Owl have been seen in the area.

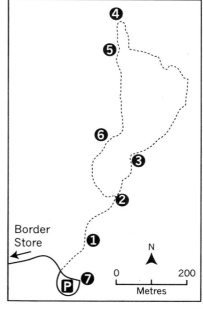

3.04 Bardedjilidji Sandstone Walk

The Bardedjilidji Sandstone Walk is well signposted off the Ubirr (Oenpelli) Road, 35 km north of the Arnhem Highway, 2 km from the Border Store. The walk passes through layered sandstone outliers adjacent to the East Alligator River. An interpretive trail has been laid out and a map for this, with numbered waypoints, is available at the start of the walk. Part of the 2.5 km track becomes flooded during the Wet Season and a somewhat shorter Wet Season Walk is marked. If sandstone birds are the main target species at the site, it is just as well to do the marked Wet Season Walk only. It is a very easy walk and particularly pleasant during the early part of the morning or late in the afternoon. However, there is little shade, so it can be very hot during the middle of the day.

The Rock-holes Walk (6 km long) starts 1 km into the Bardedjilidji Walk and passes through riparian woodland at the edge of the East Alligator River and Catfish Creek. There is a camping ground located at Merl, 3 km from Bardedjilidji.

Key Species

Grey Goshawk, Pacific Baza, Chestnut-quilled Rock-Pigeon, White-lined Honeyeater, Bar-breasted Honeyeater, Sandstone Shrike-thrush

Other Species

Pied Heron, Black-necked Stork, Helmeted Friarbird, White-throated Honeyeater, Spangled Drongo, Great Bowerbird

Finding Birds

White-lined Honeyeater may be seen in any of the flowering trees at the base of the sandstone outcrops. Chestnut-quilled Rock-Pigeon and Sandstone Shrike-thrush are most likely in the area between the numbered 4 and 6 signposts [2 - 3], though anywhere that offers views onto the top of the sandstone ridges is a good place to sit, watch and wait. Chestnut-quilled Rock-Pigeon may be seen flying between outliers or perched in shade near the top of the ridges. Listen for the characteristic whirring of

wings or the low-pitched "coo-caroo" call at any time of day. In the late afternoon, the Rock-Pigeon descends lower, coming to the ground at the edge of the woodland to feed. The area to the right before the 'gap' leading to marker 6 [3] is particularly productive at this time of day. The sandstone outcrops on the western side of the track between [4] and [5] are better later in the afternoon, when they are viewed from the shady side. Towards the end of the loop [1] there are excellent views of the higher points of some of the outcrops, so there are good chances of seeing the Rock-Pigeon or Sandstone Shrike-thrush a little earlier in the afternoon, as they rest higher up in crevices. Watch for raptors throughout the walk, as Brown Goshawk is common and Grey Goshawk and Pacific Baza are occasionally seen.

Along the Rock-holes walk, look for flowering *Melaleucas* (Paperbark trees) that grow adjacent to the East Alligator River and Catfish Creek, as Bar-breasted Honey-eater can often be found feeding in these trees. Northern Rosella and waterbirds such as Pied Heron, Black-necked Stork and Intermediate Egret are often encountered feeding next to Catfish Creek.

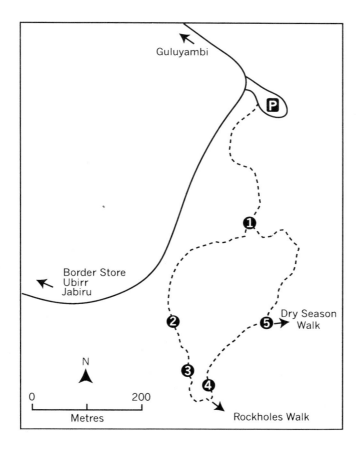

3.05 Nourlangie

Nourlangie, which is open between 8 am and sunset, is 16 km east of the Kakadu Highway, where the well signposted turn-off is 21 km south of Jabiru or 128 km north of the Park's southern border. There is an extensive sandstone rock escarpment here with pockets of monsoon forest. It is one of the best places to see Banded Fruit-Dove and White-lined Honeyeater. Visit early in the morning or in the late afternoon, when temperatures are lower. Early morning has the additional benefit of being quieter, as there will be fewer tour groups. Nawurlandja is a sandstone plateau nearby and adjacent to Anbangbang Billabong, which can be a good place to see waterbirds later in the Dry Season.

Nearest camping is at Muirella Park, 7 km south of the Nourlangie turn-off.

Key Species

Chestnut-backed Button-quail, Banded Fruit-Dove, Partridge Pigeon, Chestnut-quilled Rock-Pigeon, Emerald Dove, Black-tailed Treecreeper, White-lined Honeyeater, Helmeted Friarbird, Sandstone Shrike-thrush, Variegated Fairy-wren

Other Species

Red-tailed Black-Cockatoo, Weebill, Leaden Flycatcher, Little Woodswallow, Great Bowerbird

Finding Birds

Partridge Pigeon is often seen along the road from the Kakadu Highway to Nourlangie, particularly early or late in the day. Chestnut-backed Button-quail has been seen around the edges of the carpark [P], early in the morning before the area becomes busy. Black-tailed Treecreeper and Long-tailed and Masked Finches are generally easy to find in the woodland around the carpark or along the outer tracks [1, 3, 8]. Emerald Dove is often found sheltering in the shade of rocks at [2].

Look in shaded fig trees near the main rock art gallery [5] for Banded Fruit-Dove. Also try the small area of monsoon forest around the footbridge [6], or walk up the track a little further and sit on the seats, watching and listening for the Fruit-Dove's characteristic low-pitched 'woom-woom' call. White-lined Honeyeater may also be seen on the fringe of the monsoon forest here. The Gunwarddehwarde Lookout [7] offers a good view of the immediate area and Banded Fruit-Dove can sometimes be seen from there. The short loop walk to Anbangbang Shelter [4] is another good area to check for Banded Fruit-Dove, with a chance of also seeing White-lined Honeyeater and Sandstone Shrike-thrush. Chestnut-quilled Rock-Pigeon may be seen or heard almost anywhere on the rocky escarpment, though generally it stays quite high up on the escarpment at this location.

The nearby Anbangbang Billabong can hold large numbers of waterbirds during the latter part of the Dry Season. Green Pygmy-goose, Magpie Goose, Wandering and Plumed Whistling-Ducks and Pied Heron are among the many species that may be seen here.

Nawurlandja is a large sandstone outcrop adjacent to the Billabong. Chestnut-quilled Rock-Pigeon and Sandstone Shrike-thrush are frequently seen here. It is a short but steep and very exposed walk, so it is best to do it in the early morning or late afternoon.

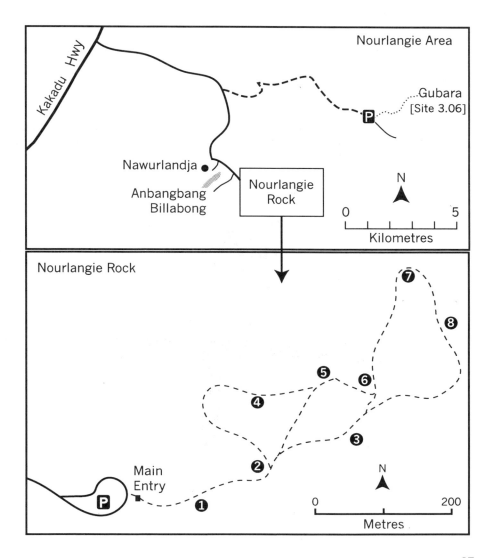

3.06 Gubara

Gubara is located 9 km from the Nourlangie Road, with a well signposted turn-off just outside the Nourlangie entrance gate (see Nourlangie Area map, previous page). The road is unsealed but generally in good condition and quite suitable for conventional vehicles. The final part of the road can be rough and depending on the current conditions it may be necessary to park at [P1], a little short of the main designated parking area [P2], but adding only a couple of minutes to the walk. Gubara itself is approximately 3.5 km walk from the carpark. It is a particularly good site for Banded Fruit-Dove and White-lined Honeyeater.

Key Species
Banded Fruit-Dove, Chestnut-quilled Rock-Pigeon, Northern Rosella, Little Bronze-Cuckoo, Barking Owl, Rainbow Pitta, Green-backed Gerygone, White-lined Honeyeater, Sandstone Shrike-thrush, Cicadabird, Little Woodswallow

Other Species
Brown Quail, Peregrine Falcon, Red-tailed Black-Cockatoo, Blue-winged Kookaburra, Azure Kingfisher, Rainbow Bee-eater, Silver-crowned Friarbird, Rufous-throated Honeyeater, Yellow Oriole

Finding Birds

The first 2 km of the walk to Gubara passes through open woodland where commonly seen birds include Red-tailed Black Cockatoo, Northern Rosella, Silver-crowned Friarbird and Rufous-throated Honeyeater. Brown Quail is often seen in the denser grassland that occurs near the creek lines.

Approximately 2 km into the walk, the track goes close to a large section of escarpment country, between [1] and [2]. Chestnut-quilled Rock-Pigeon, Sandstone Shrike-thrush and White-lined Honeyeater may be seen on the escarpment here and Peregrine Falcon and Little Woodswallow are often seen flying on the thermals above. Access to the escarpment is limited, so it is worthwhile sitting and watching from the track for bird activity. Further along, the track passes a huge fig tree [3] where Barking Owl has been seen at roost.

Approximately 3 km from the car park the track crosses a creek [4]. The log bridge that spans the creek is quite solid but take care in crossing. White-lined Honeyeater can often be found in the flowering trees around the creek.

Banded Fruit-Dove may be seen feeding or resting in the fig trees that grow along-side the stream [5]. The Fruit-Dove may still be present here even if the swimming hole is crowded, but generally it is disturbed by the noise, so try further along the creek or in the trees up the slope to the right. The fig trees around the creek are tall and it is worth taking special care to search them thoroughly, as Banded Fruit-Dove is notoriously difficult to see when perched quietly in the foliage.

Rainbow Pitta and Cicadabird may be seen in the monsoon forest patches that surround the creek and Azure Kingfisher is often seen along the creek.

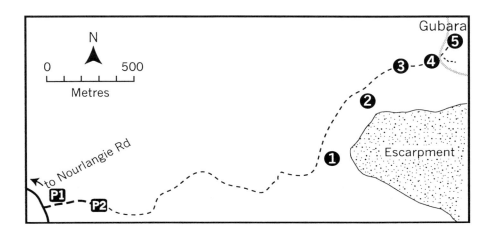

3.07 Cooinda (Yellow Water)

Cooinda is one of the few tourist developments in Kakadu, with fuel, food and accommodation. It is well signposted along the Kakadu Highway 49 km south of the Arnhem Highway and about 100 km north of the Park's southern border. Yellow Water itself is at the confluence of Jim-Jim Creek and the South Alligator River and comprises a series of billabongs surrounded by treeless floodplain and extensive woodland. Commercial boat cruises operate regularly throughout the day, providing excellent opportunities for seeing Great-billed Heron and Little Kingfisher. The early morning cruise is generally the best for birds. It is very popular so it is advisable to book ahead. Some of the guides on the boat tours are particularly helpful. Let them know you are after specific birds and they may make a special effort. A good variety of woodland birds, including Partridge Pigeon, can be found around Cooinda.

Key Species

Great-billed Heron, Black Bittern, Red Goshawk, Grey Goshawk, White-browed Crake, Chestnut-backed Button-quail, Partridge Pigeon, Little Kingfisher, Azure Kingfisher, Oriental Cuckoo, Little Bronze-Cuckoo, Brush Cuckoo, Spotted Nightjar, Australian Owlet-nightjar, White-browed Robin, Rufous Fantail, Broad-billed Flycatcher

Other Species

Magpie Goose, Radjah Shelduck, Green Pygmy-goose, Australian Pelican, Black-necked Stork, Pied Heron, Nankeen Night Heron, Intermediate Egret, Glossy Ibis, Royal Spoonbill, White-bellied Sea-Eagle, Bush Stone-curlew, Comb-crested Jacana, Gull-billed Tern, Whiskered Tern, Little Corella, Forest Kingfisher, Sacred Kingfisher, Blue-winged Kookaburra, Blue-faced Honeyeater, Bar-breasted Honeyeater, Rufous-banded Honeyeater, Shining Flycatcher, Lemon-bellied Flycatcher, Yellow Oriole

Finding Birds

Brolgas are often present on the grassy area [2] near the carpark. Black Bittern, Great-billed Heron and Azure and Little Kingfishers are often seen during an early morning cruise on Yellow Water. Plumed and Wandering Whistling-Ducks, Magpie Goose, Green Pygmy-goose, Nankeen Night-Heron, Black-necked Stork, Glossy Ibis, Australian Pelican, Comb-crested Jacana and Shining Flycatcher are also commonly encountered on the cruise. White-bellied Sea-Eagle nests around the billabong and Red, Grey and Brown Goshawks have all been seen nearby.

Pheasant Coucal, Brush Cuckoo, Little Bronze-Cuckoo, Forest and Sacred Kingfishers, Blue-winged Kookaburra, Bar-breasted, Banded and Rufous-throated Honeyeaters, Shining Flycatcher and Yellow Oriole are frequently seen in paperbark trees by the walkway [1] at the water's edge. From November to April, Oriental Cuckoo is present here too. White-browed Crake and Azure Kingfisher can sometimes be seen at the water's edge.

Australian Hobby and Grey Goshawk are often seen around the resort as both nest nearby. Chestnut-backed Button-quail has been seen around the airstrip. There is no public access to the airstrip from the resort but walking into the woodland from the Cooinda Road [4] should provide similar opportunities to see this species. Partridge Pigeon occurs quite commonly in the woodland here and is often seen along road-sides, particularly in the early morning or late afternoon.

Barking Owl is often heard at night and can easily be tracked down, feeding around the lights of the resort [3]. Bush Stone-curlew is heard regularly around the resort and can be seen on the resort lawns or in nearby open areas. A night-time drive along the Cooinda Road or Kakadu Highway can be productive. Australian Owlet-nightjar frequently sits on the road (unfortunately many become casualties as a result) and Spotted Nightjar is common in the area.

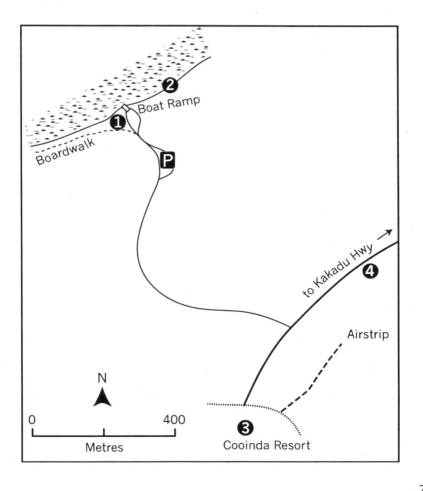

3.08 Mardugal

Mardugal is located 2 km south of Cooinda turn-off, 98 km north of the Park's southern border and 51 km south of the Arnhem Highway. There are two walking tracks here. One is a woodland walk and the other in part follows the edge of Mardugal Billabong, a neighbouring billabong to Yellow Waters. The Billabong walk may be flooded during the Wet Season and can remain closed to walkers until as late as May. There are two camping areas at Mardugal, both with excellent facilities. Electricity generators are permitted in camping area 2.

Key Species
Great-billed Heron, Black Bittern, Little Bronze-Cuckoo, Azure Kingfisher, Little Kingfisher, Green-backed Gerygone, Rufous Fantail, White-browed Robin, Little Shrike-thrush, Varied Triller

Other Species
Red-tailed Black-Cockatoo, Forest Kingfisher, Brush Cuckoo, Dollarbird, Dusky Honeyeater, White-throated Honeyeater, Bar-breasted Honeyeater, Banded Honeyeater, Shining Flycatcher, Leaden Flycatcher, Lemon-bellied Flycatcher, Northern Fantail, White-bellied Cuckoo-Shrike, Spangled Drongo, Olive-backed Oriole, Yellow Oriole, Great Bowerbird, Long-tailed Finch

Finding Birds
In the early morning or late afternoon Great-billed Heron can often be seen from the bridge where the Kakadu Highway crosses Jim-Jim Creek [4]. Partridge Pigeon is frequently seen along the edge of the highway or around Camping Area 1 [1].

The dense vegetation around the Mardugal Billabong boat ramp [3] and carpark area is good for Little Shrike-thrush, Green-backed Gerygone, White-browed Robin, Shining Flycatcher, Northern Fantail, Rufous Fantail, Leaden Flycatcher and Dusky Honeyeater. Little and Azure Kingfishers and Black Bittern may be seen perched on the lower branches of the thick vegetation at the billabong's edges near the boat ramp. Bar-breasted and Banded Honeyeaters can often be seen in the Paperbark trees around the carpark. Forest Kingfisher, Little Bronze-Cuckoo, White-bellied Cuckoo-Shrike, Grey Shrike-thrush, Long-tailed Finch and Great Bowerbird may be seen in the woodlands to the rear of the Boat Ramp carpark.

Mardugal Billabong walk starts from camping area 2 and passes through open woodland where Varied Triller, Forest Kingfisher, Spangled Drongo, Dollarbird and Lemon-bellied Flycatcher can be seen. Where the track meets and follows the billabong, look in the low riverine vegetation for Azure and Little Kingfishers and White-browed Robin. This is also a good area for Great-billed Heron. The Billabong track is not a loop walk and after some distance it becomes indistinct. At this point there is little choice but to turn and take the same route back.

Gun-gardun walk, approximately 2 km in length, is a loop that passes through woodland. Black-tailed Treecreeper is found throughout the walk. When trees are in flower there can be large numbers of Varied Lorikeets. Red-winged Parrot, Blue-winged Kookaburra and Olive-backed Oriole are usually quite easily seen here and Rufous-throated and Banded Honeyeaters are often present along with the commoner Brown, White-throated and White-gaped Honeyeaters. There are a few areas where the woodland gives way to more open grassland and finches and mannikins can be expected in these areas.

At night, Bush Stone-curlew, Barking Owl, Spotted Nightjar and Australian Owlet-nightjar can often be heard. A spotlighting walk or drive along the road between the camping grounds and the Kakadu Highway can be a good way to see these species.

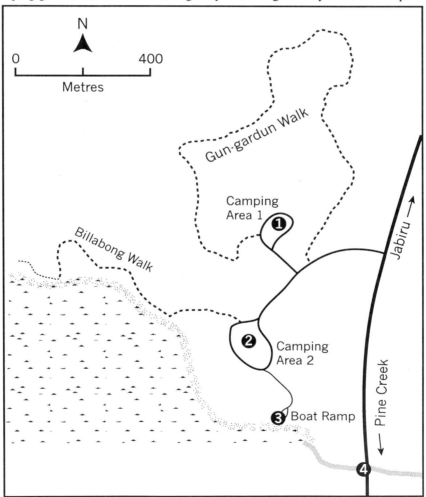

3.09 Plum Tree Creek

Plum Tree Creek crosses the Gunlom Road (see next site for location details) at two points. From the Kakadu Highway, only the second crossing [2] is signposted, approximately 24 km east of the turn-off from the Kakadu Highway and 15 km from Gunlom. This area is an alternative site to Gunlom for the White-throated Grasswren and does not require a strenuous walk up an escarpment ridge, but as the site has been irregularly covered its reliability for the Grasswren is uncertain. There are no tracks in this area, so care must be taken with walking as well as with navigating. The walk is not especially strenuous but much of it is exposed and there is some scrambling over rocky outcrops. It is advisable to wear strong footwear and a hat and to carry plenty of water. There is a good camping area at Gunlom, about 15 km further north, and basic camping at Kambolgie Creek, 8 km back toward the Kakadu Highway. Mary River Roadhouse, which is on the Kakadu Highway just south of the National Park border, has motel and cabin accommodation.

Key Species
Red Goshawk, Chestnut-quilled Rock-Pigeon, White-throated Grasswren, White-lined Honeyeater, Sandstone Shrike-thrush, Gouldian Finch

Other Species
Northern Rosella, Red-tailed Black-Cockatoo, Banded Honeyeater, Red-backed Fairy-wren, Weebill, Leaden Flycatcher, Crimson Finch, Double-barred Finch, Long-tailed Finch, Masked Finch, Great Bowerbird

Finding Birds
Access to the sandstone outcrops is easier from near the northern creek crossing. Park off the road near the crossing [2], then walk due west toward the highest outcrops [5, 6]. The vegetation near the roadside is dense, with no clear tracks, but once through this, the sandstone areas are much more open and quite easy to walk in. Chestnut-quilled Rock-Pigeon and Sandstone Shrike-thrush can sometimes be seen at the first major line of sandstone [4]. It can be worth the effort even coming this far, then turning down the gully toward the creek [7] and walking back to the road. Beyond [4] the terrain gradually gets higher. Around the high 'peaks' [5, 6] the slopes are densely covered with spinifex, low scrub and large sandstone boulders. This is the prime habitat of White-throated Grasswren and it is here that chances of seeing the Grasswren will be greatest. Find a good vantage point and sit quietly, scanning the suitable habitat below and listening for the Grasswren's calls. Typically this species is most active early in the morning as the sunlight starts to fall on the sandstone boulders, though it may be seen throughout the day. Chestnut-quilled Rock-Pigeon and Sandstone Shrike-thrush also occur here. These species are likely to be observed atop the sandstone boulders or in the sheltered cracks or gaps between boulders.
It is preferable to return the same way, or along the creek from [7], rather than try and walk due south to the road, as the terrain becomes difficult in that direction.
Within the escarpment area there are numerous creek lines with thick scrubby

vegetation and scattered *Eucalyptus* trees at their edges. White-lined Honeyeater can often be found in these areas.

Birdwatching anywhere along Plum Tree Creek itself can be productive, though normally the best section is between [3] and [7]. Red-backed Fairy-wren, Double-barred Finch and Crimson Finch are common in the long grass anywhere along the creek and Black Bittern is also often seen perched in trees beside or overhanging the creek. Watch along the creek and surrounding area for Red Goshawk, which has been seen here. Finches, including Gouldian, come to drink at the remnant pools of water along the creek during the latter part of the Dry Season.

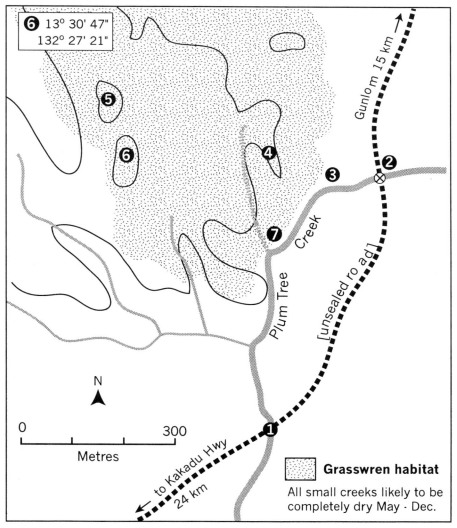

3.10 Gunlom (Waterfall Creek)

Gunlom is 39 km east of the Kakadu Highway. The turn-off is well signposted 138 km south of the Arnhem Highway or 11 km north of the Park's southern border. The spinifex covered escarpment country above Gunlom is the best place to see White-throated Grasswren. It is worth asking the campground manager about recent reports of the Grasswren as birdwatchers often tell the manager about their sightings when staying there. In addition to the well maintained camping facilities, there is a cold shower for day visitors. This can be refreshing after a hot climb up the escarpment. Note that the road from the Kakadu Highway to Gunlom is unsealed. For much of the Wet Season it is closed even to four-wheel drive vehicles and after most other roads in the Park have reopened this one may remain closed. From November until at least May, it is essential to check. This can be done at the entrance stations, Park Headquarters or Mary River Roadhouse (or see *Further Information,* p160).

Key Species
Black Bittern, Pacific Baza, Square-tailed Kite, Rose-crowned Fruit-Dove, Banded Fruit-Dove, Chestnut-quilled Rock-Pigeon, Barking Owl, Rufous Owl, Rainbow Pitta, White-throated Grasswren, Variegated Fairy-wren, Helmeted Friarbird, White-lined Honeyeater, White-browed Robin, Sandstone Shrike-thrush

Other Species
Red-backed Fairy-wren, Banded Honeyeater, Rufous-throated Honeyeater, Shining Flycatcher, Little Woodswallow, Great Bowerbird

Finding Birds
White-throated Grasswren occurs where there are large clumps of spinifex amongst boulders on the slopes and valleys of the escarpment. Climb to the top of the escarpment via the well marked path [1]. The Grasswren has been seen immediately at the top on both sides [2, 3] of the waterfall. It has often been seen on the escarpment up to the right [4], but it is easier walking and generally more successful to aim for the lower outcrops further on. Follow the track above the creek for a couple of hundred metres, keeping to the edge of the escarpment. After the track turns from rock to sand, the high, layered boulders on the right start to level out [5]. Walk up the gentle slope on the right towards the layered sandstone outcrop [6] about 150 m from the creek. Grasswrens are frequently seen on or from this outcrop and a good way to locate them is simply to wait and listen for the call, which could be at any time of day. Once a call is heard, it is quite easy to track down the birds. Grasswrens come down the slopes, presumably to drink, a number of times during the day in the Dry Season, so observations at any time are possible. However, it is still best to get to the site quite early in the morning, even though Grasswren activity may not occur until some time later. Chestnut-quilled Rock-Pigeon and White-lined Honeyeater can be found anywhere that the Grasswren may be seen, while the Sandstone Shrike-thrush is more likely to be seen higher up on the rocky outcrops [7]. The Banded Fruit-Dove prefers the fig trees that grow next to the stream though it may also be seen in the small

patches of monsoon forest higher up on the escarpment. Look out also for Pacific Baza, Square-tailed Kite, Little Woodswallow, the sandstone race of Helmeted Friarbird and Rufous-throated and Banded Honeyeaters.

Near the base of the waterfall [8] it is possible to follow the creek through an area of monsoon forest [9]. Watch along here for Black Bittern, Pacific Baza, Barking Owl, Rufous Owl, Rose-crowned Fruit-Dove, Emerald Dove, White-browed Robin, Shining Flycatcher and Rainbow Pitta.

Northern Rosella, Barking Owl, Bush Stone-curlew, Banded Honeyeater, Masked Finch, Partridge Pigeon and Great Bowerbird are frequently seen in and around the Gunlom campground [10, 11].

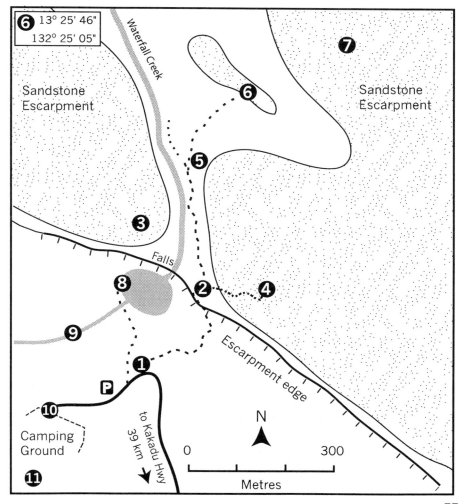

3.11 Bukbukluk

Bukbukluk Lookout is clearly signposted, approximately 130 km south of the Arnhem Highway and 19 km north of the southern Park border. Bukbukluk Creek is 1.7 km north of the Lookout.

Key Species
Brown Quail, Pacific Baza, Grey Goshawk, Barking Owl, Banded Fruit-Dove, Rose-crowned Fruit-Dove, Hooded Parrot, Northern Rosella, Oriental Cuckoo, White-browed Robin, Bar-breasted Honeyeater, Hooded Robin

Other Species
Little Corella, Red-tailed Black-Cockatoo, Common Koel, Common Bronzewing, Blue-faced Honeyeater, Rufous-throated Honeyeater, Spangled Drongo

Finding Birds
There is a short walk to the Lookout [1] from the carpark [P]. Northern Rosella and Red-winged Parrot are frequently seen in the woodland here and Hooded Parrot has been recorded. Black-tailed Treecreeper is often present near the top of the lookout.

Hooded Parrot, Northern Rosella, Common Bronzewing and Hooded Robin have been seen in the roadside woodland between the Lookout and the Creek, most often in the hilly area [2]. There is no designated parking area at the creek but there is sufficient space to park near the bridge. Banded Fruit-Dove and Rose-crowned Fruit-Dove feed and rest in the larger fig trees that grow along the creek. White-browed Robin occurs in the denser vegetation along the creek. Pacific Baza, Grey Goshawk, Barking Owl, Oriental Cuckoo and Common Koel have also been seen along the creek. Brown Quail are often in the thick grassland near the edge of the riverine vegetation. When the woodland beside the creek is in flower, Varied Lorikeet and Banded, Rufous-throated and Bar-breasted Honeyeaters can usually be seen here.

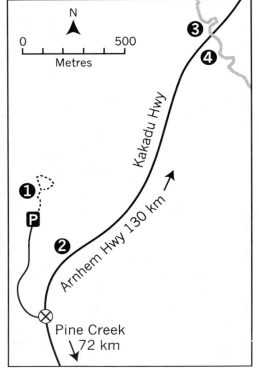

78

4 Katherine Region

The Katherine Region covers the area within approximately 100 km radius of the township of Katherine, which is 300 km south of Darwin. About 240 bird species have been recorded in the Katherine Region, including the hard to find Red Goshawk, Chestnut-backed Button-quail and Crested Shrike-tit.

The shops and services in Katherine (the third largest centre in the Northern Territory, behind Darwin and Alice Springs) are adequate to meet most travellers' needs. It is a good place to stock up with supplies if travelling to other more isolated areas such as Kakadu National Park or the southwest Top End.

There is hotel accommodation and camping in Pine Creek, Katherine and Mataranka and camping in Nitmiluk and Elsey National Parks.

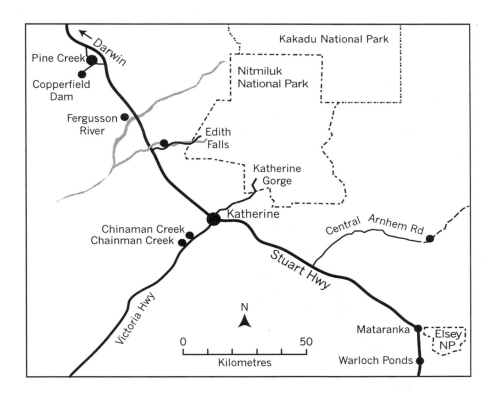

4.01 Pine Creek

Pine Creek is just off the Stuart Highway, approximately 90 km from Katherine and 210 km from Darwin. It is a great birding area in its own right but is also convenient as a place to stay before or after a trip into Kakadu National Park, since it is only about 60 km south of the Park. An early morning visit to Copperfield Dam (see site 4.02) or to Fergusson River (see site 4.03) is easy from here. Camping, caravan and hotel accommodation are available in the town.

Key Species
Black Falcon, Partridge Pigeon, Diamond Dove, Hooded Parrot, Northern Rosella, Black-chinned Honeyeater, Gouldian Finch

Other Species
Wedge-tailed Eagle, Crested Pigeon, Red-tailed Black-Cockatoo, Cockatiel, Red-winged Parrot, Blue-faced Honeyeater, Great Bowerbird, Masked Finch, Long-tailed Finch

Finding Birds
Hooded Parrot has been seen in many places around the town area, mainly in the late afternoon when birds come in to drink and to rest in the shade of the trees. One of the few places where birds are frequently seen throughout the day is in the trees near the top of the well signposted Mine Lookout road [1], or along the walking track adjacent to this road. The water-tower road [2] is also worth a try for these parrots. The tops of both these roads give views over the township and surrounding area and can be excellent viewing areas for raptors including Wedge-tailed Eagle and Black Falcon.

Hooded Parrot may also be found in the park [3] opposite the hotel, the shady trees around the Pine Creek Police Station [4], the trees and power lines along the Water

Gardens [5] or around the Racetrack [8]. They have occasionally been recorded in woodland around the sewage ponds [7]. The parrots may sit quietly in trees in these areas in the late afternoon and early morning, so it can be a good option to drive slowly around the township in the late afternoon, stopping occasionally to look and listen for them.

Red-winged Parrot, Cockatiel, Crested Pigeon, Blue-faced Honeyeater and Great Bowerbird are generally common in and around the township and Masked and Long-tailed Finches regularly drink at the taps and sprinklers in the Water Gardens [4].

The nearby sewage ponds [6] are accessed from the clearly marked cemetery road from the Stuart Highway. While the ponds themselves generally hold only small numbers of the commoner waterbirds such as Australasian Grebe, Grey Teal, Black-fronted Dotterel or Black-winged Stilt, the woodland [7] around the sewage works can be very productive. Gouldian Finch, Diamond Dove, Partridge Pigeon, Black-chinned Honeyeater and Hooded Parrot have been recorded here.

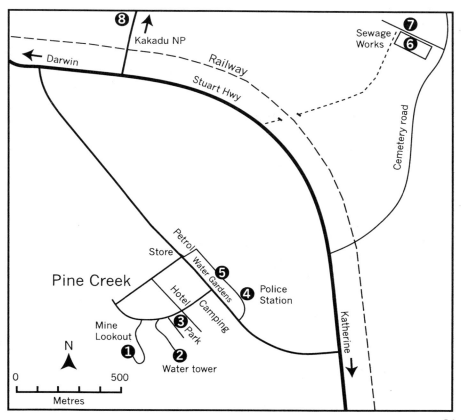

4.02 Copperfield Dam

Copperfield Dam is located approximately 5 km from the Stuart Highway, along the Umbrawarra Gorge road, which is about 2 km south of Pine Creek. The road is gravel but it is generally well maintained and is suitable for conventional vehicle. There are camping and picnic lunch facilities and the Dam offers swimming all year. As well as the specific sites mentioned in the text below it can be worthwhile walking some way around the edge of the Dam.

Key Species
Chestnut-backed Button-quail, Partridge Pigeon, Diamond Dove, Cockatiel, Hooded Parrot, Northern Rosella, Black-tailed Treecreeper, Gouldian Finch

Other Species
Brown Quail, White-faced Heron, Australian Darter, Red-tailed Black-Cockatoo, Little Corella, Sulphur-crested Cockatoo, Rainbow Lorikeet, Varied Lorikeet, Pallid Cuckoo, Brush Cuckoo, Forest Kingfisher, Silver-crowned Friarbird, Little Friarbird, Red-backed Fairy-wren, Grey Shrike-Thrush, White-winged Triller, Pied Butcherbird, Leaden Flycatcher, Restless Flycatcher, Rufous Whistler, Great Bowerbird, Long-tailed Finch, Masked Finch

Finding Birds
Chestnut-backed Button-quail occurs in grassy woodland throughout the Copperfield Dam area. It has most frequently been found on the grassy slopes between track (6) and the Umbrawarra Gorge road [9]. The Button-quail has also been seen in the woodland around the northern track [4], approximately 500m north of the campground, just to the west of a small hill [3]. Pallid and Brush Cuckoos, Forest Kingfisher, Northern Rosella, Restless Flycatcher, Black-tailed Treecreeper, Silver-crowned and Little Friarbirds and Masked and Long-tailed Finches are also frequently seen here. Red-backed Fairy-wren can be easily seen in the vicinity of the Go-Kart Track entrance.

The camping area [2] is also very good for birds. Rainbow and Varied Lorikeets are commonly found in flowering *Eucalyptus* trees around the campground and Northern Rosella occurs here regularly. Partridge Pigeon can sometimes be seen around the campground early in the morning and Chestnut-backed Button-quail has also been seen here. Some of the commoner waterbirds of the Top End including Masked Lapwing, White-faced Heron, Little Pied Cormorant and Australian Darter may be seen on or around the Dam itself. Black-tailed Treecreeper may be seen in the woodland along tracks [4] and [5]. This species may be seen on the trunks of trees or on termite mounds and is generally very vocal so is easy to track down. Watch for Black-chinned Honeyeater along track [6]. Diamond Dove, Partridge Pigeon and Long-tailed and Masked Finches are also seen along here. Brown Quail is commonly seen in the lower area [7] below track [6].

Hooded Parrot and Partridge Pigeon may be seen in the woodland along the Umbrawarra Gorge Road. Northern Rosella, Red-tailed Black-Cockatoo, Rufous Whistler and Leaden Flycatcher are also frequently seen in this woodland.

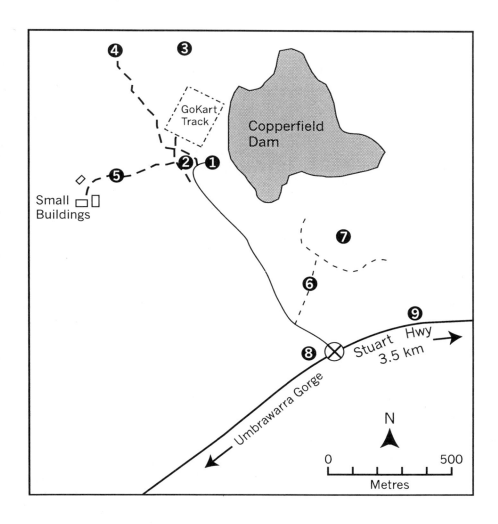

4.03 Fergusson River

The Fergusson River crosses the Stuart Highway about 60 km northwest of Katherine, or 35 km southeast of Pine Creek. The river provides a good drinking area for finches (most notably Gouldian) and Hooded Parrot, especially from about June to November, when the surrounding area is very dry. It is best to visit this site around dawn.

Key Species

Black Bittern, Black Falcon, Square-tailed Kite, Diamond Dove, Hooded Parrot, Cockatiel, Northern Rosella, Black-tailed Treecreeper, White-throated Gerygone, Yellow-tinted Honeyeater, Rufous-throated Honeyeater, Black-chinned Honeyeater, Banded Honeyeater, Hooded Robin, Jacky Winter, Gouldian Finch

Other Species

Whistling Kite, Black Kite, Rainbow Lorikeet, Little Corella, Blue-winged Kookaburra, Red-backed Kingfisher, Blue-faced Honeyeater, Pied Butcherbird, Double-barred Finch, Masked Finch, Long-tailed Finch

Finding Birds

The access track, which is rather bumpy, is about 60 m from the northwestern end of the bridge. Avoid the maze of smaller tracks and park near the loop [1], just before the track dips steeply down. Follow the track down to the river bank and along the river to the west, just short of where the river bends to the right [2]. Hooded Parrots are sometimes perched in the trees here and may be located by their high-pitched tinkling call. Hooded Parrot, Gouldian, Long-tailed, Masked and Double-barred Finches and Banded, Rufous-throated and occasionally Black-chinned Honeyeaters, come to drink at the gravelly edge of the river at [3]. Check the trees along the bank here also, as these species often perch in them before flying down to drink.

An easy walk along the bank towards the bridge, or further, is usually worthwhile. A range of other birds including Black Bittern, Nankeen Night Heron, Blue-faced Honeyeater, Banded Honeyeater, Jacky Winter and Striated Pardalote may be seen near the water. Diamond Dove, Cockatiel and White-throated Gerygone are often present in the drier woodland nearby. Some of the scarcer birds of prey such as Black Falcon and Square-tailed Kite have been seen here quite frequently.

The area south of the river [inset] can be productive in the early morning. Many species, including Gouldian Finch and Hooded Parrot, drink at the river and feed in the nearby woodland. A good place to try is around the wayside stop, about 3 km south of the river [4]. Gouldian Finch may be found feeding in the grass on the side of the road here. Hooded Parrot, Northern Rosella, Black-tailed Treecreeper, Varied Sittella, Jacky Winter and Hooded Robin are more likely to be seen in the trees on the western side of the road [5]. There are a few indistinct tracks on this side and it is worth walking along some of these to get a little further into the woodland, increasing the chances of seeing some of these birds.

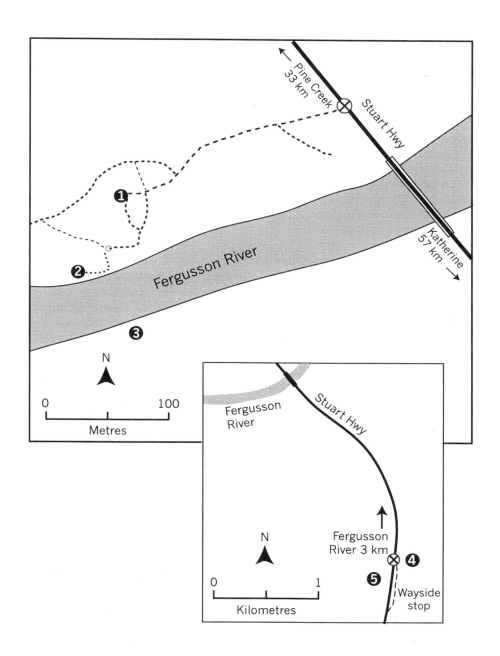

Pine Creek 33 km

Stuart Hwy

Fergusson River

Katherine 57 km

❶

❷

❸

N

0   100
Metres

Fergusson River

Stuart Hwy

N

Fergusson River 3 km

❹

❺

Wayside stop

0   1
Kilometres

4.04 Edith Falls Road

Edith Falls Road runs off the Stuart Highway, about 45 km south of Pine Creek and 40 km from Katherine. Almost anywhere along the road provides opportunities to see Gouldian Finch and Hooded Parrot; however, the area around the creek, 5 km in from the Stuart Highway, has been found to be quite reliable. The Falls are at the end of the road in Nitmiluk National Park, where camping and kiosk facilities are available.

Key Species
Black Bittern, Hooded Parrot, Bar-breasted Honeyeater, Gouldian Finch

Other Species
Red-tailed Black-Cockatoo, Red-winged Parrot, Northern Rosella, Brush Cuckoo, Azure Kingfisher, Banded Honeyeater, Grey Shrike-thrush, Jacky Winter, Great Bowerbird, Crimson Finch, Masked Finch, Long-tailed Finch

Finding Birds
Take the small track that leaves the road just past the creek, about 5 km along from the Stuart Highway. Depending on the height of the grass, access to the creek may or may not be straightforward. Area [2] has extensive edges that allow birds to come in to drink. In dry years, when the creek is only a series of small pools, many birds including Hooded Parrot and Gouldian Finch come to drink here. Black Bittern and Azure Kingfisher are sometimes seen perched in dense vegetation along the creek, particularly between [1] and [5]. Woodland birds such as Grey Shrike-thrush, Jacky Winter, Long-tailed Finch and Masked Finch are often seen to the north of the track [3]. The grass on the slope [4] to the west of the track provides a feeding area for Gouldian Finch.

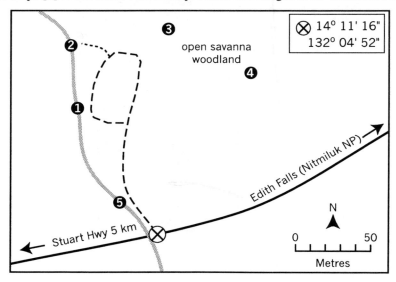

4.05 Katherine Sewage Ponds

The Katherine Sewage Ponds are about 7 km southwest of the town centre. Turn off the Victoria Highway into Novis Quarry Road, 5.5 km from the Stuart Highway. There is a sign-posted 'birdwatchers turnstile' but this is likely to be locked and overgrown with vegetation. However, birds within the sewage works can generally be viewed satisfactorily from the fence with the aid of a spotting scope. The sewage ponds have shown potential but there have not yet been any significant reports of birds that are not easily seen elsewhere. Nevertheless, these sewage works are worth monitoring and so the site is included here.

Key Species
Wedge-tailed Eagle, Black Falcon

Other Species
Pied Heron, Radjah Shelduck, Pheasant Coucal

Finding Birds
Park off the road near the gate [1] and walk along the fence near the turnstile [2] to get views of the ponds. The track at the side of the ponds [3] provides some access to woodland but offers no better views of the ponds than at the front. Crested Pigeon, Varied Lorikeet, Pheasant Coucal and other common woodland birds may be seen in the surrounding woodland [3, 4]. Check the creek crossings [5, 6], as birds come to drink at the remnant pools in the late afternoon. The rubbish tip attracts numerous raptors including Wedge-tailed Eagle and Black Falcon, which sometimes perch in woodland [7] opposite the tip during the hotter part of the day.

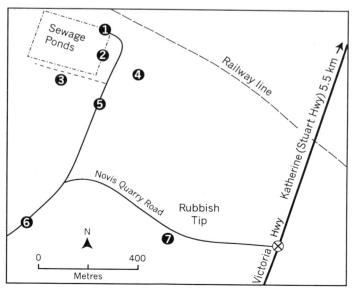

4.06 Chinaman Creek

Chinaman Creek is approximately 16 km west of Katherine. Take the dirt track just west of the Manballoo Airfield sign, about 200 m east of the actual creek crossing. There are some rough patches on the track but it is generally suitable for conventional vehicles.

Key Species
Chestnut-backed Button-quail, Hooded Parrot, White-throated Gerygone, Black-chinned Honeyeater, White-browed Woodswallow, Masked Woodswallow, Black-tailed Treecreeper, Gouldian Finch

Other Species
Red-winged Parrot, Cockatiel, Weebill, Rufous-throated Honeyeater, Rufous Whistler, Grey-crowned Babbler, Masked Finch, Long-tailed Finch

Finding Birds
Chestnut-backed Button-quail, Gouldian Finch and Hooded Parrot may come to drink at the remnant pools of water, mainly [1] but also [2], in the early morning or late afternoon. White-throated Gerygone, Black-tailed Treecreeper, Rufous-throated, Banded and Black-chinned Honeyeaters occur in the woodland around the old Victoria Highway [3]. Black-faced and Little Woodswallows occur here regularly and White-browed and Masked Woodswallows occur occasionally.

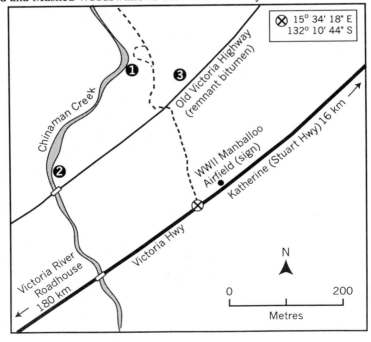

4.07 Chainman Creek

Chainman Creek crosses the Victoria Highway about 20 km from Katherine. Its main attraction is that the savanna woodland nearby is good for Chestnut-backed Button-quail. Hooded Parrot, Gouldian Finch and Crested Shrike-tit also occur in the area.

Key Species

Black-breasted Buzzard, Chestnut-backed Button-quail, Northern Rosella, Hooded Parrot, White-throated Gerygone, Black-chinned Honeyeater, Black-tailed Treecreeper, Varied Sittella, Crested Shrike-tit, Gouldian Finch

Other Species

Wedge-tailed Eagle, Red-winged Parrot, Cockatiel, Weebill, Rufous-throated Honeyeater, Rufous Whistler, Masked Finch, Long-tailed Finch

Finding Birds

Chestnut-backed Button-quail occur on the savanna woodland slopes around the creek. Park off the road at the crest of the hill (1) about 800m along the highway southwest of the creek. The road here has small but noticeable upward sloping banks at its edges. Walk the short distance to the fence through the woodland. The grass here can be quite tall but it is not particularly dense. The best area to find the Button-quail is generally in the longest grass, so look for this and try there first. The grassy area continues in all directions but the most successful area for finding the Button-quail is around the crest of the hill [shaded area, 2 -3]. The fence is not continuous, so try also on the northern side of the fence line or along the old highway [4]. The Button-quail flush easily but with care it is possible to see birds on the ground.

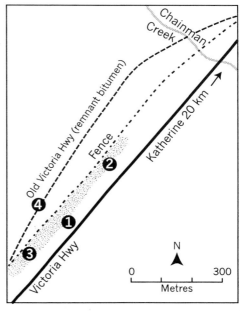

Hooded Parrot has been seen in the woodland here and there are many reports of Gouldian Finch at nearby waterholes and woodland. Northern Rosella, Black-tailed Treecreeper and Varied Sittella occur regularly.

Watch out for the rare Crested Shrike-tit, as it has been reported from the woodland in this area.

89

4.08 Mataranka (Elsey National Park)

Mataranka is on the Stuart Highway about 105 km south of Katherine. There are two access points into Elsey National Park, located at either end of the town. The Northern entrance to the park is via Martin Road and is well signposted from the township. Entry to the southern section, which includes the popular Thermal Pool, is well signposted from the Stuart Highway just south of Mataranka. There are camping facilities in the Park and a range of accommodation options in and around the town.

Key Species
Great-billed Heron, Black Bittern, Red Goshawk, Grey Goshawk, Black-breasted Buzzard, Hooded Parrot, Oriental Cuckoo, Barking Owl, Rufous Owl, White-browed Robin, Black-chinned Honeyeater, Singing Honeyeater, Grey-fronted Honeyeater

Other Species
Wedge-tailed Eagle, Varied Lorikeet, Sacred Kingfisher, Yellow-throated Miner, Banded Honeyeater, Rufous-throated Honeyeater, Dusky Honeyeater, Yellow-tinted Honeyeater, Bar-Breasted Honeyeater, Shining Flycatcher, Apostlebird, Great Bowerbird

Finding Birds
The northern section of the Park (Bitter Springs) is entered via Martin Road. Numerous raptors, mainly Black Kite but often also Black-breasted Buzzard and Wedge-tailed Eagle, can be seen flying over the open areas along this road. When in flower, the paperbark trees near the car park attract Rainbow and Varied Lorikeets as well as Banded, Bar-breasted, Rufous-throated, Dusky and Yellow-tinted Honeyeaters. Many of the commoner birds can be seen along the pleasant loop walk [1] around the spring, though the woodland and river edge to the north of the entry road is much more productive. A few rather indistinct tracks go into the woodland and paperbark swamp here. None provides a clear path as the vegetation near the river is dense but it is reasonably easy to navigate a path roughly parallel to the river [2 - 3]. Red, Grey and Brown Goshawks have all been seen here. Black Bittern, Azure Kingfisher and Shining Flycatcher occur in the vegetation along the river bank [3] and Crimson Finch and Red-backed Fairy-wren can usually be seen in the grassland near the river. Hooded Parrot is sometimes present in the adjacent woodland [4].

The southern entrance of the Park goes to Mataranka Thermal Pool. Great Bowerbird and Apostlebird are tame around the caravan park [5]. Both Rufous and Barking Owls also occur here, the latter sometimes seen hunting moths under lights in the quiet parts of the campground. An early morning or evening walk along the track from Mataranka Thermal Pool to Stevies Hole [6] will give a good chance of seeing Great-billed Heron and Black Bittern on the edge of the river.

There is easy access to the river at 4-Mile [8]. Great-billed Heron and Black Bittern may be seen here or anywhere along the river where access is possible. Unfortunately the designated walk along the river is not always well maintained and can become poorly marked and rough.

In the drier woodland that surrounds the Botanical Walk [7], Black-chinned, Grey-fronted and Singing Honeyeaters are sometimes seen along with the more regular Rufous-throated, Banded and Yellow-tinted Honeyeaters.

A night time drive along John Hauser Drive can be a good way to find Australian Owlet-nightjar and Spotted Nightjar.

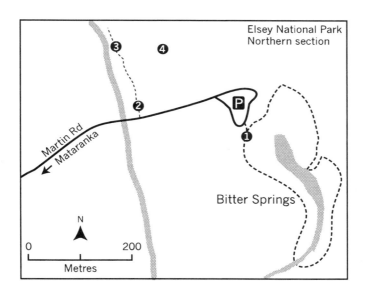

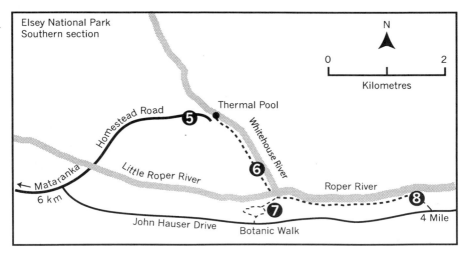

4.09 Central Arnhem Road

The Central Arnhem Road runs off the Stuart Highway about 45 km south of Katherine. Birdwatching can be very productive anywhere along this road, so stop and check any incidental bird activity. The road passes over some small creeks and these can be good places to see finches drinking, particularly late in the year (Aug - Nov). Hooded Parrot, Gouldian Finch and Pictorella Mannikin can occur almost anywhere in the woodlands and Crested Shrike-tit is a rare possibility. The actual site described here is 50 km from the Stuart Highway and a little over 2 km beyond the currently sealed section of the Central Arnhem Road. The Central Arnhem Road continues to the coast but a permit is required to travel into Arnhem Land, about 180 km beyond this site.

Key Species
Hooded Parrot, Crested Shrike-tit, Red-browed Pardalote, Pictorella Mannikin, Gouldian Finch

Other Species
Northern Rosella, Cockatiel, Varied Lorikeet, Black-tailed Treecreeper, Yellow-tinted Honeyeater, Black-chinned Honeyeater, Little Woodswallow, Masked Finch

Finding Birds
Stop by the sign indicating 50 km to the Stuart Highway [1]. The woodland here is quite open and appears to be a favoured area for parrots. The grass is not very dense and comes right down to the road. Masked and Long-tailed Finches and sometimes Pictorella Mannikin can be seen directly at the roadside as a result of this. Shortly beyond the end of the bitumen a creek crosses the road and remnant Dry Season pools provide excellent drinking sites for parrots, honeyeaters and finches. Gouldian Finch can be expected here and Crested Shrike-tit has been recorded in the area. Red-browed Pardalote can occur throughout the woodland or along the creek and is easily detected by its distinctive call.

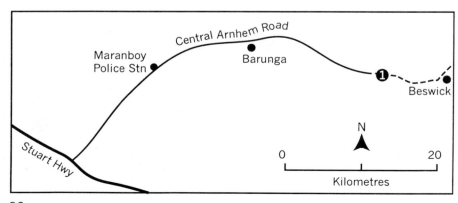

4.10 Warloch Ponds

Warloch Ponds is a small, semi-permanent wetland area that crosses the Stuart Highway, approximately 30 km south of Mataranka. For our purposes it marks the southern edge of the Top End. For birdwatchers travelling up from the arid southern regions it may be the first place they see some of the typical Top End waterbirds such as Black-necked Stork, Pied Heron and Comb-crested Jacana.

Key Species
Australian Bustard, Hooded Parrot, Crested Shrike-tit, Gouldian Finch, Pictorella Mannikin

Other Species
Diamond Dove, Common Bronzewing, Black-tailed Treecreeper, Banded Honeyeater, Yellow-tinted Honeyeater, White-plumed Honeyeater, Masked Finch, Long-tailed Finch

Finding Birds
The wetland can be viewed from the Stuart Highway but there is only a small and narrow area for parking, so it may be best to park at the Borrow Pits track [1] and walk to the bridge. The derelict bridge [3] on the Elsey Cemetery road provides an alternative view of the wetland. In general, the waterbirds are found on the eastern side of the Stuart Highway, where the water is deeper and more extensive. Waterbird numbers increase as the day progresses, so an afternoon visit is best for this. Australian Bustard (late in the Dry Season), Diamond Dove, Common Bronzewing, Hooded Parrot, Cockatiel and a range of finch and honeyeater species come to drink, mainly on the western side [4], where there are more extensive shallow edges. A track [1] on the southern side of the bridge gives access to the woodland [2] where Crested Shrike-tit has been seen. This species can sometimes be heard ripping bark from the paperbark trees when feeding. Crested Shrike-tit has also been seen in the woodland along Gorrie Road. Try the area about 5 km in from the Stuart Highway [5]. Black-tailed Treecreeper also occurs here and Hooded Parrot, Gouldian Finch and Pictorella Mannikin are occasionally seen in the area.

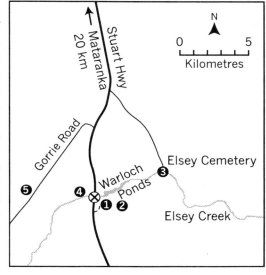

93

5 Southwest Region

The Southwest Region covers the area along the Victoria Highway from the Buntine Highway, 125 km west of Katherine, to Keep River National Park, on the Western Australian border. A little over 230 species, including White-quilled Rock-Pigeon and Purple-crowned Fairy-wren, have been recorded in the region. The region is renowned for its finches, with eleven species recorded. In addition to this, there are many other birds here that are either absent or rare elsewhere in the Top End. These include Spinifex Pigeon, Grey-fronted Honeyeater, Red-capped Robin, Crimson Chat and Pictorella Mannikin.

The Victoria Highway can be a haven for raptors. Grey Falcon, Square-tailed Kite, Black-breasted Buzzard and Spotted Harrier are reported more regularly from here than elsewhere in the Top End.

In addition to the specific sites described here, stop at creek crossings, particularly when water levels are low, as birds come to drink at the remnant pools at all times of day. Stopping to look in areas where incidental bird activity is observed can also result in finding some of the nomadic birds that are normally difficult to locate.

There are camping facilities at Victoria River, Timber Creek and within Gregory and Keep River National Parks. There are hotel and fuel facilities in Timber Creek and Victoria River.

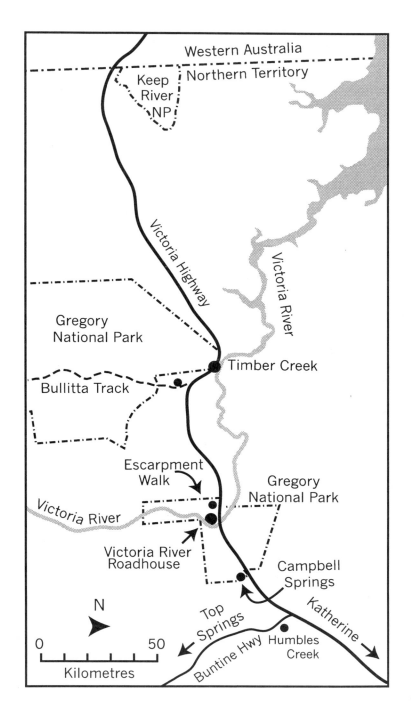

Western Australia

Northern Territory

Keep River NP

Victoria Highway

Victoria River

Gregory National Park

Bullitta Track

Timber Creek

Escarpment Walk

Gregory National Park

Victoria River

Victoria River Roadhouse

Campbell Springs

Top Springs

Katherine

N

0   50
Kilometres

Buntine Hwy   Humbles Creek

5.01 Humbles Creek & Buntine Highway

The Buntine Highway runs south from the Victoria Highway, about 125 km west of Katherine. There are records of Crimson Chat, Budgerigar and other inland birds along this road, so it can be well worth travelling 20 - 30 km along here and stopping frequently to check the roadside woodland. Humbles Creek, 8 km south of the Victoria Highway, can be very good for birding particularly during the latter Dry Season months, from August to November.

Key Species
Square-tailed Kite, Grey Falcon, Budgerigar, Red-browed Pardalote, Star Finch, Crimson Chat

Other Species
Cockatiel, Diamond Dove, Horsfield's Bronze-Cuckoo, Apostlebird, Zebra Finch

Finding Birds
Star Finch occurs along Humbles Creek and can sometimes be seen in the grass or creekside vegetation [1 - 2]. Red-browed Pardalote is often present in the *Eucalyptus* trees [2] by the creek or in the nearby woodland. Apostlebirds are attracted to water in the cattle yard [4], as are Cockatiel, Diamond Dove, finches and honeyeaters. Raptors including Spotted Harrier, Black-breasted Buzzard, Square-tailed Kite and Grey Falcon occur regularly in the area during the Dry Season and may be seen around the creek or along the Buntine Highway.

On Delamere Station, about 30 km along the Buntine Highway from the Victoria Highway, there is quite a large lake that attracts numerous waterbirds. The lake is not visible from the road, although there is sometimes overflow water in nearby low-lying areas along the roadside. The lake is on private property but it is well worth bird-watching in the nearby roadside woodland because of the birds attracted to the lake. Grey Falcon hunts at the lake and is sometimes seen soaring over the adjacent woodland. Australian Bustard, Budgerigar and Crimson Chat have been seen in the roadside woodland between the lake and the Victoria Highway.

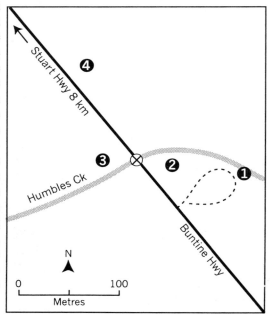

5.02 Campbell Springs

Campbell Springs is about 155 km west of Katherine, 32 km west of the Buntine Highway or 38 km east of Victoria River crossing. The small dirt track is marked by a signpost, though not a large one. The springs provide permanent water and so the site can be very productive particularly from July to November.

Key Species
Diamond Dove, Red-browed Pardalote, Gouldian Finch, Yellow-rumped Mannikin, Star Finch, Masked Finch, Long-tailed Finch

Other Species
Cockatiel, Rainbow Bee-eater, Red-backed Fairy-wren, Jacky Winter, Crimson Finch, White-winged Triller, Golden-headed Cisticola

Finding Birds
Cockatiel and Diamond Dove are often seen along the entry track or around the carpark [P]. Red-backed Fairy-wren and Golden-headed Cisticola can usually be found in the tall grass along the track to the creek [1]. Red-browed Pardalote is often present in trees along or near the creek and can be readily located by its distinctive call. Crimson Finch can be seen in the *Pandanus* by the creek [1] and Masked and Long-tailed Finches come to drink at the rocky edges close by. Gouldian Finch and Yellow-rumped Mannikin tend to drink further downstream. Cross the creek at [1] and follow it west through what may be rather tall grass for about 150 m, until the creek opens out again and there is a large rocky platform [2]. The finches often land in the scrubby trees by the creek before venturing down to the edges of the rock platforms to drink.

From about June to September, depending on availability of seeding grasses, quite large groups of Gouldian Finch may be seen throughout the day feeding on the side of the highway [3].

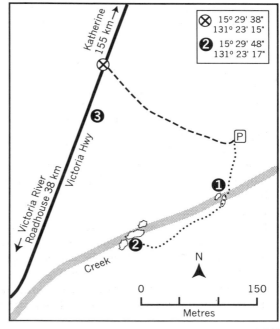

97

5.03 Victoria River

The Victoria River crosses the Victoria Highway 194 km SW of Katherine, or 91 km E of Timber Creek. Along its sides are large areas of savanna woodland and sandstone escarpment. The bridge is one of the easiest places to see Purple-crowned Fairy-wren, and one of the closer places to Darwin for both Yellow-rumped Mannikin and Star Finch. The Escarpment Walk is exposed and invariably hot and it is essential to carry plenty of water when doing this walk. Note that for convenience, the labels on the inset map of the Escarpment Walk coincide with the numbered markers along the walk.

Key Species
Grey Falcon, White-quilled Rock-Pigeon, Purple-crowned Fairy-wren, Variegated Fairy-wren, Ground Cuckoo-shrike, Sandstone Shrike-thrush, Yellow-rumped Mannikin, Star Finch

Other Species
Wedge-tailed Eagle, White-bellied Sea-Eagle, Osprey, Whistling Kite, Red-tailed Black-Cockatoo, Varied Lorikeet, Barking Owl, Dollarbird, Rainbow Bee-eater, Striated Pardalote, Banded Honeyeater, Bar-breasted Honeyeater, Yellow-tinted Honeyeater, Chestnut-breasted Mannikin, Crimson Finch, Double-barred Finch

Finding Birds
Ground Cuckoo-shrike is uncommon throughout the Top End, but is quite frequently seen around the Gregory National Park entry, to the east of the river.

Purple-crowned Fairy-wren occurs in the tall dense canegrass along the Victoria River. It has been seen at all corners of the bridge [9] and at all times of day, so it is usually just a matter of patience. Yellow-rumped Mannikin and Star Finch are frequently seen in the same place, particularly when the grass is seeding. Both Grey and Black Falcon have been seen quite regularly in the area around the river and Barking Owl is frequently seen in the caravan park.

A good alternative site for Purple-crowned Fairy-wren, particularly early in the morning, is further up the river. Follow the road south at the Victoria River Access sign and park at the end of the bitumen [10]. Walk along the sandy track [11] towards the river, watching and listening for the Fairy-wren on both sides of the track. Look in the area around the top of the sandy track for Variegated Fairy-wren, which is frequently seen here.

Escarpment Walk
Star Finch may be present in quite large numbers if the grassland that surrounds the carpark [P] is seeding. The Walk passes over two sections of escarpment between the track markers [3] and [4] and between track markers [4] and [5] before reaching the top of the Plateau. White-quilled Rock-Pigeon may be seen sitting on the rock-shelves in these escarpment areas. They are sometimes flushed off the track itself, normally during the heat of the day when they sit under the shade of thicker vegetation. At the

top of the second escarpment, between track markers [4] and [5] the track passes close to the edge of the ridge giving a great view of the entire escarpment edge. This is a good place to sit and watch as Rock-Pigeons may be seen flying between exposed ridges or perched in shade near the top of the ridges. Listen for the characteristic whirring of wings or the low-pitched "coo-caroo" call.

Sandstone Shrike-thrush may be seen on top of the plateau between markers [5] and [8]. This species is often very vocal in the early morning or late afternoon and thus easier to locate at those times.

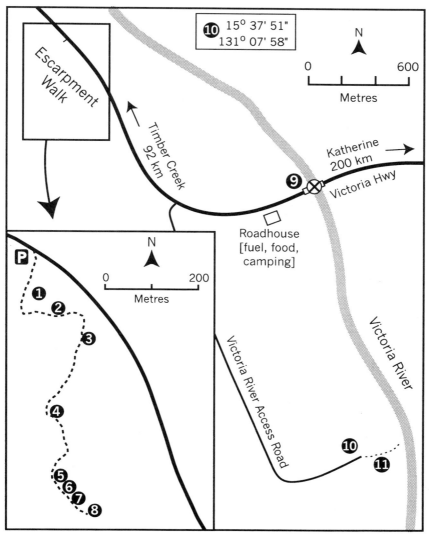

5.04 Bullita Access Road

The Bullita Access Road runs south from the Victoria Highway, 11 km east of Timber Creek, 80 km west of Victoria River and approximately 275 km west of Katherine. The roadside provides excellent opportunistic birding. It is possible to travel by conventional vehicle as far as Bullita Homestead and Limestone Gorge, provided the road is in good condition. However, birding can be excellent within a few kilometres of the highway, so it is not necessary to travel far.

Key Species
Black-breasted Buzzard, Square-tailed Kite, Grey Falcon, Spinifex Pigeon, Red-browed Pardalote, Grey-fronted Honeyeater, Grey-headed Honeyeater, Crimson Chat, Pictorella Mannikin, Yellow-rumped Mannikin, Star Finch, Gouldian Finch

Other Species
Diamond Dove, Banded Honeyeater, Rufous-throated Honeyeater, Singing Honeyeater, Little Woodswallow, Long-tailed Finch, Masked Finch

Finding Birds
Red-browed Pardalote can be found along or near the creek quite close to the small causeway [1]. Its distinctive call is frequently heard, though the bird itself can be difficult to locate. Banded, Grey-fronted and Black-chinned Honeyeaters may be seen in the woodland along the road [2]. Look in the roadside woodland for flowering trees, which attract these and other honeyeaters and Varied Lorikeet.

Spinifex Pigeon may be seen anywhere along the road. It is usually easiest to find in the late afternoon when it is more likely to come right to the road edge. Pictorella and Yellow-rumped Mannikins and Gouldian Finch may also be seen anywhere along the roadside. Watch out for activity in grassy areas, as this will invariably be finches and

mannikins and there is a good chance that any such group will comprise more than one species. Crimson Chat, which is rarely recorded in the Top End, has been seen in the woodland in this area. Watch for raptors, as Square-tailed Kite, Black-breasted Buzzard and Grey Falcon have been seen here, along with the more common raptors. Grey-headed Honeyeater, rare in the Top End, has been seen a little further south along this road.

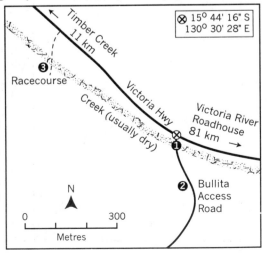

5.05 Timber Creek

Timber Creek is about 285 km west of Katherine. It is an excellent place for finches and mannikins, with ten species occurring regularly. The Airfield is about 6 km from the township.

Key Species

Black Bittern, Boobook Owl, Barking Owl, Crimson Chat, White-browed Robin, Gouldian Finch, Yellow-rumped Mannikin, Star Finch, Painted Finch

Other Species

Varied Lorikeet, Diamond Dove, Rufous-throated Honeyeater, Banded Honeyeater, Yellow-tinted Honeyeater, Long-tailed Finch, Masked Finch, Zebra Finch

Finding Birds

Several honeyeater species, along with a range of finches, including Gouldian, come to drink at the taps and sprinklers on the town's nature strip [1] and the lawns of the town's camping grounds [2, 3]. Often these birds come in from the woodland on the opposite side of the road [4], so it is worthwhile checking that area as well. White-browed Robin is often seen in the more densely vegetated areas along the fence or creek at the rear of the campgrounds. Look in the vegetation along the creek for Black Bittern also, as it may be seen perched on branches beside or overhanging the water. At night, Boobook and Barking Owls may be seen hunting around the street lights in front of the shops or hotel.

At the airfield, Gouldian and Star Finches and Yellow-rumped Mannikin are often seen in the early morning or late afternoon in the short grass along either side of the main airfield fence. Spinifex Pigeon and Singing Bushlark are frequently seen here and Crimson Chat and Painted Finch have also been recorded here but are rare.

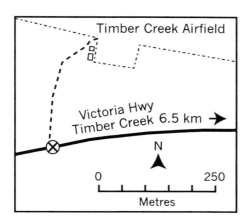

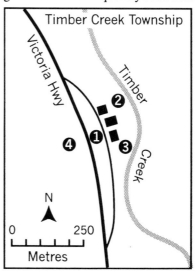

5.06 Victoria River Access

A little over 4 km northwest of the Timber Creek township a gravel track on the east side of the highway gives access to the Victoria River. The track is normally suitable for a conventional vehicle. Birding is very productive near the river, particularly early in the morning, when finches and honeyeaters tend to concentrate here. The open woodland and grassland along the track are also well worth checking, either early in the morning or towards evening.

Key Species

Black-breasted Buzzard, Grey Falcon, Spinifex Pigeon, Diamond Dove, Purple-crowned Fairy-wren, Red-browed Pardalote, Grey-fronted Honeyeater, Black-chinned Honeyeater, Gouldian Finch, Pictorella Mannikin, Yellow-rumped Mannikin, Star Finch

Other Species

Red-winged Parrot, Varied Lorikeet, Rufous-throated Honeyeater, Banded Honeyeater, Yellow-tinted Honeyeater, Long-tailed Finch, Masked Finch, Zebra Finch

Finding Birds

Watch along the sides of the main track for Spinifex Pigeon, particularly on the higher ground [1]. Grey-fronted Honeyeater, Black-tailed Treecreeper, Varied Sittella and Bush Stone-curlew may be seen along the northern track [2–4]. Follow the main track to its end and walk to the edge of the cliff [3] overlooking the river. Gouldian and Star Finches and Yellow-rumped Mannikin may be seen feeding in the grass, or perched in the small trees growing here. Purple-crowned Fairy-wren is often present in the tall dense canegrass along the banks of the river here or further upstream [4]. Brown Quail, Bush Stone-Curlew, Spinifex Pigeon, Diamond Dove, Black-tailed Treecreeper, Varied Sittella and Grey-fronted Honeyeater may be seen in the woodland along the track to [4]. Grey Falcon is sometimes also seen in this area and Black-breasted Buzzard is recorded commonly.

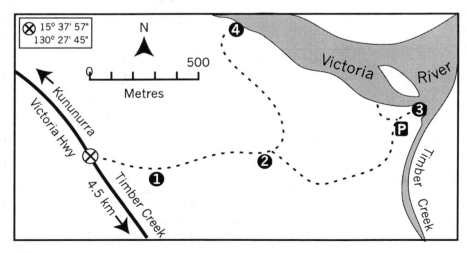

5.07 Western Victoria Highway

Many areas along the Victoria Highway between Timber Creek and Keep River National Park (or the Western Australian border) provide excellent birdwatching opportunities. This section of the Top End is somewhat more arid than the remainder of the region covered in this book and several species that are rarely seen elsewhere in the Top End can be seen more easily here. Late in the Dry Season, large numbers of birds come to drink at remnant pools in the creeks and rivers, and it is worth spending time in the vicinity of any pools where bird activity is seen, regardless of the time of day. Many woodland birds including Emu, Australian Bustard, Spinifex Pigeon and a range of passerine species may be seen feeding by the roadside in the early morning or late in the afternoon. Some of the scarcer raptors such as Grey Falcon and Square-tailed Kite occur regularly in the area.

Key Species
Emu, Australian Bustard, Black-breasted Buzzard, Little Eagle, Spotted Harrier, Square-tailed Kite, Grey Falcon, Budgerigar, Red-browed Pardalote, Grey-fronted Honeyeater, Crimson Chat, Ground Cuckoo-shrike, Gouldian Finch, Star Finch, Yellow-rumped Mannikin, Pictorella Mannikin

Other Species
Diamond Dove, Pallid Cuckoo, Cockatiel, Northern Rosella, Red-backed Kingfisher, Jacky Winter, Black-tailed Treecreeper, Masked Woodswallow, White-browed Woodswallow, Little Woodswallow, Zebra Finch, Brown Songlark

Finding Birds
Other than incidental roadside birding, there are a few specific sites worth targeting. Between 20 and 30 km west of Timber Creek there are two small wetland areas [1] that retain water well into the Dry Season and attract moderate numbers of waterbirds along with a range of doves, honeyeaters and finches. The East Baines River [2], approximately 60 km west of Timber Creek, is generally a reliable site for Red-browed Pardalote. Dingo Creek [3], a little west of the Duncan Highway turn-off, is a regular Dry Season drinking site for Gouldian Finch.

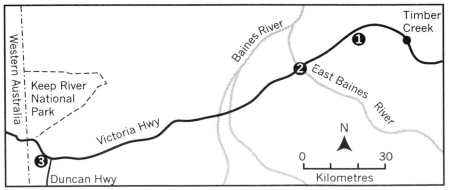

5.08 Keep River National Park

Keep River National Park is 468 km west of Katherine, at the border between the Northern Territory and Western Australia. There are two widely separated campgrounds, both with only basic facilities. While the range of birds around both areas is similar, White-quilled Rock-Pigeon is more common on the escarpment in the northern (Jarrnarm) area.

Key Species
Emu, Square-tailed Kite, Black-breasted Buzzard, Peregrine Falcon, Grey Falcon, Spinifex Pigeon, White-quilled Rock-Pigeon, Budgerigar, Grey-fronted Honeyeater, Sandstone Shrike-thrush, Pictorella Mannikin, Gouldian Finch, Star Finch

Other Species
Plumed Whistling-Duck, Green Pygmy-goose, Little Eagle, Wedge-tailed Eagle, Spotted Harrier, Diamond Dove, Cockatiel, Singing Honeyeater, Silver-crowned Friarbird, Rufous-throated Honeyeater, Great Bowerbird, Australian Magpie, Zebra Finch

Finding Birds
Cockatoo Lagoon [1] attracts a large range of waterbirds, particularly later in the Dry Season, from about August onward. Finches including Star Finch and Yellow-rumped Mannikin can often be seen around the lagoon's grassy edges. Watch along the main Park road for Spinifex Pigeon, which occurs throughout the area and for Emu, which is generally scarce in the Top End. Budgerigar is an irregular visitor and may occur anywhere but is most likely to be seen flying fast overhead in small groups.

From the Gurrandalng campground [2] there is a marked walk through some rocky escarpment. In the morning or late afternoon, Sandstone Shrike-thrush may be seen atop the rocky outcrops. White-quilled Rock Pigeon is generally more secretive, so watch the shaded areas under the rocky overhangs. Peregrine Falcon is frequently seen soaring over the higher escarpment areas. Spotted Harrier hunts over the lower grassy areas surrounding the escarpment and both Square-tailed Kite and Black-breasted Buzzard occur in the open woodland areas.

A short distance from the Gurrandalng campground the road crosses a small creek. For much of the Dry Season this is no more than a few small pools but it serves as an important water source for a range of birds in the surrounding area. Gouldian, Masked and Long-tailed Finches and sometimes Pictorella Mannikin can be seen drinking at these pools, along with Grey-fronted and other honeyeater species. Grey Falcon is occasionally also seen in this area, as it preys on the birds that come to drink.

From the Jarrnarm campground [3] the track marked 'Western Walk' gives close access to some spectacular escarpment and White-quilled Rock Pigeon is generally easy to find here. The walk passes through open woodland and grassland before getting to the escarpment edge. Watch for the Rock-Pigeon as the track reaches the escarpment [4 - 5]. It is quite common here and very close views are possible if the birds are not unduly disturbed.

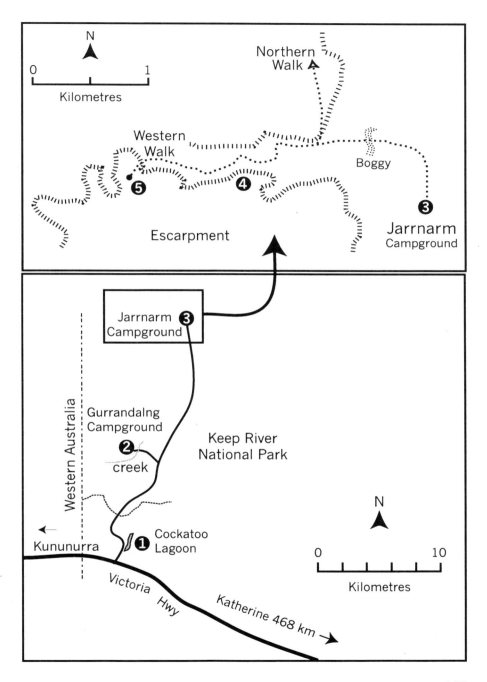

N

0   1
Kilometres

Northern
Walk

Western
Walk

Boggy

5

4

3

Jarrnarm
Campground

Escarpment

Jarrnarm **3**
Campground

Gurrandalng
Campground

2

creek

Keep River
National Park

Western Australia

1 Cockatoo
Lagoon

Kununurra

Victoria Hwy

Katherine 468 km →

N

0   10
Kilometres

Annotated List of Birds of the Top End

The annotated list is a summary of the abundance, distribution, seasonality of occurrence and habitat of birds occurring in the Top End (see p.6 for area covered). Numbers are given in brackets [ ] to indicate the birdwatching sites (chapters 1 to 5) where a species may be more easily seen, or more regularly reported, but not every site where the species may occur is listed. Sites are not generally given for species that are common and widespread, or for species that are so rare as to make site lists misleading. Some commoner species may be listed with fewer sites than uncommon species, since more of the sites where an uncommon species has been recorded may need to be visited to give the best chance of finding these species.

Abundance

The terms 'common', 'moderately common' and 'rare' are used to indicate how easily (or otherwise) a species may be seen in the Top End. These terms are merely for guidance, based on the authors' experiences and have no statistical basis.

Seasonality

'Wet Season' is used loosely for the period October - April and 'Dry Season' for the period May - September. In many cases however, actual months are given, as the two-season system is not sufficiently precise for the seasonality of these species. A summary of the Top End climate appears elsewhere in this book (p. 10). Note that many birds are nomadic, with irregular movement into or within the Top End (see p. 9).

Distribution

Distribution is given mainly in terms of the five regions into which the authors have divided the area covered by the book. Where the terms 'southern' or 'northern' Top End are used, the township of Pine Creek may be considered to be at the dividing line.

Habitat

Top End habitats are briefly described elsewhere in this book (p. 7).

National rarities are considered to be confirmed through acceptance by the Birds Australia Rarities Committee (BARC). Reports of species that are rare in the Northern Territory only, are considered to be confirmed through vetting by the authors. Published reports of rare species that have not been confirmed are included for completeness and are indicated as 'unconfirmed'. Case numbers are provided for those claims that have been assessed by BARC.

Bird names and the taxonomic order used throughout this book follow the standard for Australian birds set out by Christides and Boles (1994).

Emu *Dromaius novaehollandiae*
[2.04 5.03 5.07 5.08]
Uncommon to rare resident, perhaps declining. Recorded in all regions, though mainly in southern and eastern Top End. Open woodland, plains, river valleys. Most commonly seen along and south of the Victoria Highway.

Orange-footed Scrubfowl *Megapodius reinwardt*
[1.04 1.05 1.07 1.09 1.12 1.16 2.01 3.03 3.10]
Common coastal and near-coastal resident, recorded in all regions except Katherine. Monsoon forest and adjacent mangroves, parks, urban areas.

Brown Quail *Coturnix ypsilophora*
[1.03 1.09 1.10 1.17 3.02 3.05 3.06 3.07 3.11 4.02 5.02 5.03 5.05 5.06 5.08]
Common and widespread resident, recorded in all regions. Woodlands, floodplains, wetland edges.

King Quail *Coturnix chinensis*
[1.12 1.13 3.01 5.08]
Rare to uncommon coastal and near coastal resident. Wetland edges, floodplains.

Magpie Goose *Anseranas semipalmata*
[1.04 1.05 1.10 1.12 1.13 1.17 2.01 2.03 3.01 3.02 3.07 5.08]
Common and widespread resident, recorded in all regions. In Darwin area, numbers greatest September - December. Most wetland habitats, orchards, well watered parks.

Plumed Whistling-Duck *Dendrocygna eytoni*
[1.10 1.13 2.03 2.06 3.01 3.02 3.07 5.08]
Common and widespread resident, recorded in all regions. Usually less coastal in distribution than the following species. Wetlands including farm dams, sewage ponds.

Wandering Whistling-Duck *Dendrocygna arcuata*
[1.10 1.13 1.17 2.03 2.06 3.01 3.02 3.07 5.08]
Common and widespread resident, recorded in all regions. Most wetland habitats including sewage ponds.

Blue-billed Duck *Oxyura australis*
Vagrant. Single record, Leanyer Sewage Works, March 1985 (Noske & van Gessel 1987).

Black Swan *Cygnus atratus*
Rare, irregular visitor, mainly May - November. All regions. Most records are from southern areas. Deeper wetlands.

Radjah Shelduck *Tadorna radjah*
[1.04 1.09 1.10 1.13 1.17 2.01 2.03 3.02 3.07]
Common and widespread resident, recorded in all regions. Coasts, harbours, estuaries, most wetland habitats including sewage ponds, parks.

Pink-eared Duck *Malacorhynchus membranaceus*
[1.10 1.13 1.17 4.05 5.08]
Moderately common and widespread visitor, recorded in all regions. Present mainly April - November, but subject to yearly fluctuations depending on climatic conditions. Wetlands including sewage ponds.

Green Pygmy-goose *Nettapus pulchellus*
[1.10 1.13 1.17 2.01 2.03 2.06 3.02 3.07 5.08]
Common and widespread resident, recorded in all regions. Wetlands, occasionally sewage ponds.

Australian Wood Duck *Chenonetta jubata*
[1.10 1.13 3.02]
Rare, irregular visitor, with most records May - October. Recorded in Darwin, Fogg Dam, Kakadu and Southwest regions. Wetlands including sewage ponds.

Pacific Black Duck *Anas superciliosa*
Common and widespread resident, recorded in all regions. Greatest numbers May - December. Wetlands including sewage ponds.

Grey Teal *Anas gracilis*
Common to moderately common and widespread visitor. Recorded in all regions. Chiefly present May - December, scarce and sometimes absent in other months most years. Wetlands including sewage ponds.

Chestnut Teal *Anas castanea*
Vagrant. Recorded in Darwin region only, most recently Leanyer Sewage Works, June 1990.

Garganey *Anas querquedula*
[1.10 1.13 1.17 2.03]
Rare, irregular visitor, October - April. Recorded in Darwin, Fogg Dam and Kakadu regions. Wetlands including sewage ponds.

Hardhead *Aythya australis*
[1.10 1.13 1.17 2.03 3.02 5.08]
Moderately common to common and widespread visitor, recorded in all regions. Present mainly April - November. Wetlands including sewage ponds.

Australasian Grebe *Tachybaptus novaehollandiae*
Common and widespread resident, recorded in all regions. Greatest numbers April - November, and reduced numbers, particularly in more coastal areas, during other months. Wetland habitats including sewage ponds.

Eurasian Little Grebe *Tachybaptus ruficollis*
Vagrant. Single record, Leanyer Sewage Works, September - October 1999 [BARC 278].

Hoary-headed Grebe *Poliocephalus poliocephalus*
[1.10]
Uncommon to rare, sporadic visitor, May - October. Recorded in all regions, though mainly southern Top End. Wetlands, including sewage ponds.

Great Crested Grebe *Podiceps cristatus*
[1.10 1.18 3.07]
Uncommon to rare and localised visitor, chiefly May - October. Recorded in all regions. Deeper wetlands, dams, sewage ponds.

Tahiti Petrel *Pseudobulweria rostrata*
Rare pelagic visitor, offshore.

Streaked Shearwater *Calonectris leucomelas*
[1.08 1.09]
Uncommon to rare pelagic visitor offshore, occasionally inshore. Most records May - October.

Wedge-tailed Shearwater *Puffinus pacificus*
[1.08]
Rare pelagic visitor offshore, occasionally inshore.

Wilson's Storm-Petrel *Oceanites oceanicus*
Uncommon pelagic visitor offshore.

Red-tailed Tropicbird *Phaethon rubricauda*
Rare visitor, mainly offshore.

Brown Booby *Sula leucogaster*
[1.05 1.06 1.07 1.08 1.09]
Uncommon to moderately common coastal visitor, with most records October - April. Offshore, inshore, harbours.

Australian Darter *Anhinga novaehollandiae*
Common and widespread resident. Recorded in all regions. Inshore, harbours, estuaries, creeks, rivers, wetlands, sewage ponds.

Little Pied Cormorant *Phalacrocorax melanoleucos*
Common and widespread resident, recorded in all regions. Coasts, harbours, wetlands.

Pied Cormorant *Phalacrocorax varius*
[1.01 1.05 1.06 1.08 1.09 1.15]
Moderately common resident along coast, uncommon to rare elsewhere. Recorded in all regions. Coasts, estuaries, harbours, deeper inland waters.

Little Black Cormorant *Phalacrocorax sulcirostris*
Common and widespread resident, recorded in all regions. Coasts, harbours, wetlands.

Great Cormorant *Phalacrocorax carbo*
[1.18 2.06 5.03]
Uncommon to rare visitor, mainly May - October. Recorded in all regions. Coasts and deeper wetlands.

Australian Pelican *Pelecanus conspicillatus*
Common and widespread resident. Recorded in all regions. Coasts, harbours, wetlands, sewage ponds.

Great Frigatebird *Fregata minor*
[1.09 1.05 1.06 1.08]
Rare visitor, mainly offshore but occasionally inshore, particularly during windy weather. All year, with most records November - April.

Lesser Frigatebird *Fregata ariel*
[1.01 1.05 1.06 1.08 1.09]
Uncommon to moderately common visitor to coast. Mainly October - April. Offshore, inshore, harbours, occasionally over land in windy weather. Most frigatebirds seen in the Top End will be this species.

Christmas Frigatebird *Fregata andrewsi*
Vagrant. Recorded during periods of gale force winds. First confirmed record over Nakara (northern suburbs of Darwin) Jan 1974 (BARC 03).

Little Bittern *Ixobrychus minutus*
[1.12 2.01]
Very rare, sporadic visitor. Recorded in Darwin and Fogg Dam regions. Dense vegetation of wetlands.

Black Bittern *Ixobrychus flavicollis*
[1.10 1.12 1.14 1.16 1.17 3.04 3.07 3.08 3.10 4.03 4.04 4.08 5.05]
Uncommon and widespread resident, recorded in all regions. Densely vegetated wetlands, mangroves, riverine woodlands, paperbark forests.

Nankeen Night Heron *Nycticorax caledonicus*
[1.07 1.09 1.10 1.12 1.14 1.16 2.01 2.03 3.01 3.07 4.03]
Moderately common and widespread resident, recorded in all regions. Wetlands, paperbark forests, mangroves.

White-faced Heron *Egretta novaehollandiae*
Moderately common and widespread resident, recorded in all regions. Less common near the coast than further inland, particularly November - April. Wetlands.

Little Egret *Egretta garzetta*
Moderately common to common and widespread resident. Recorded in all regions. Estuaries, harbours, reefs, wetlands including sewage ponds.

Eastern Reef Egret *Egretta sacra*
[1.01 1.02 1.05 1.06 1.08 1.09 1.15 2.04]
Moderately common coastal resident. Recorded in Darwin, Kakadu and Southwest regions. Coasts, harbours, estuaries.

White-necked Heron *Ardea pacifica*
Moderately common and widespread resident, recorded in all regions. Less common near the coast than further inland, particularly November - April. Wetlands.

Great-billed Heron *Ardea sumatrana*
[1.01 1.02 1.09 1.15 2.04 2.06 3.04 3.07 3.08 4.08]
Uncommon to rare resident, recorded in all regions. Estuaries, harbours, large river systems and adjacent wetlands.

Pied Heron *Ardea picata*
[1.10 1.12 1.13 2.01 2.03 2.04 2.06 3.01 3.02 3.06 3.07 5.08]
Common and widespread resident, recorded in all regions. Can be uncommon to scarce in coastal areas December to February. Wetlands including sewage ponds, rubbish tips.

Great Egret *Ardea alba*
Moderately common and widespread resident. Recorded in all regions. Estuaries, harbours, wetlands.

Intermediate Egret *Ardea intermedia*
Common and widespread resident, recorded in all regions. Wetlands, well watered parks and ovals.

Cattle Egret *Ardea ibis*
[1.12 1.13 1.17 2.01 2.02 2.03 2.04]
Moderately common and widespread resident, recorded in all regions. Range contracts November - April, when breeding. Pastures with cattle, buffalo or horses, wetlands.

Striated Heron *Butorides striatus*
[1.01 1.02 1.05 1.06 1.07 1.08 1.09 1.15 2.04 3.01]
Moderately common resident. Recorded in Darwin, Kakadu and Southwest regions. Mangroves, reefs, estuaries, creeks.

Glossy Ibis *Plegadis falcinellus*
[1.04 1.11 1.12 1.13 1.17 2.01 2.03 2.04 2.06 3.01 3.02 3.07 5.08]
Moderately common and widespread visitor, with small numbers breeding. Recorded in all regions. Present in largest numbers May - October. Wetlands.

Australian White Ibis *Threskiornis molucca*
Common and widespread resident, recorded in all regions. Greatest numbers present April - November. Mangroves, wetlands including sewage ponds, parks.

Straw-necked Ibis *Threskiornis spinicollis*
Common and widespread visitor, with greatest numbers present April - November, with small numbers breeding. Recorded in all regions. Wetlands, parks, occasionally open woodland.

Royal Spoonbill *Platalea regia*
Moderately common and widespread resident, recorded in all regions. Greatest numbers April - November. Wetlands.

Yellow-billed Spoonbill *Platalea flavipes*
Rare and irregular visitor, recorded in all regions. Most records May - October. Wetlands.

Black-necked Stork *Ephippiorhynchus asiaticus*
[1.10 1.12 1.13 1.17 2.01 2.03 2.04 3.01 3.02 3.06 3.07 5.08]
Moderately common and widespread resident, recorded in all regions. Wetlands including sewage ponds, occasionally rubbish tips.

Osprey *Pandion haliaetus*
[1.01 1.02 1.03 1.05 1.06 1.07 1.08 1.09 1.15 5.03]
Moderately common resident. Recorded in all regions. Coast and adjacent areas including urban areas, large river systems.

Pacific Baza *Aviceda subcristata*
[1.06 1.08 1.09 1.12 1.15 1.16 2.01 3.03 3.04 3.10 3.11]
Uncommon but widespread resident. Recorded in all regions, though rare in Southwest. Woodlands, monsoon and riverine forest, occasionally parks and urban areas.

Square-tailed Kite *Lophoictinia isura*
[1.12 3.02 3.10 4.03 5.01 5.03 5.04 5.05 5.08]
Uncommon but widespread visitor, recorded in all regions. Open woodland, riverine woodland.

Black-breasted Buzzard *Hamirostra melanosternon*
[1.13 1.17 2.02 2.05 3.01 3.02 4.03 4.08 5.01 5.03 5.04 5.05 5.06 5.07 5.08]
Moderately common and widespread resident, recorded in all regions. More numerous in southern half of Top End, particularly in the Southwest, though numbers increase in northern areas May - October. Open woodlands, floodplains, wetland edges.

Black-shouldered Kite *Elanus axillaris*
[1.11 1.12 2.01 2.02 2.03 3.01 3.02 3.07 5.05 5.08]
Moderately common and widespread resident, recorded in all regions. Greatest numbers May - December though less seasonal variation in south than north. Grasslands, floodplains, wetland edges.

Letter-winged Kite *Elanus scriptus*
[2.02 2.04 3.01]
Rare, sporadic visitor. Occasional irruptions into the Top End, sometimes with breeding colonies. Recorded in all regions. Grasslands, floodplains, wetland edges.

Black Kite *Milvus migrans*
Common and widespread resident, recorded in all regions. Numbers much reduced throughout Top End, particularly in coastal areas, November - March. Open woodland, beaches, rubbish tips, wetland edges, sewage ponds, parks, urban areas.

Whistling Kite *Haliastur sphenurus*
Common and widespread resident, recorded in all regions. In coastal areas may appear more numerous December - April due to lower numbers of previous species during those months. Wetlands, open woodland, sewage ponds, rubbish tips, urban areas.

Brahminy Kite *Haliastur indus*
Common resident of coastal areas, recorded in all regions except Katherine. Coast, mangroves, wetlands, sewage ponds, urban areas.

White-bellied Sea-Eagle *Haliaeetus leucogaster*
Common and widespread resident, recorded in all regions. Coast, wetlands, estuaries and large river systems.

Swamp Harrier *Circus approximans*
[1.11 1.12 1.13 2.01 3.01 3.02 3.07]
Uncommon but widespread visitor, recorded in all regions. Most frequently recorded May - December. Floodplains, wetlands.

Spotted Harrier *Circus assimilis*
[1.11 1.12 2.01 2.02 3.01 3.07 5.01 5.02 5.03 5.08]
Uncommon but widespread resident, recorded in all regions. Northerly records generally restricted to May - October. Grasslands, dry floodplains, occasionally in the more open woodlands.

Grey Goshawk *Accipiter novaehollandiae*
[1.02 1.07 1.10 1.14 1.16 2.06 3.03 3.04 3.07 3.11 4.08]
Uncommon but widespread resident, recorded in all regions. Monsoon and riverine forest, mangroves, parks, urban areas.

Brown Goshawk *Accipiter fasciatus*
Common and widespread resident, recorded in all regions. Greatest numbers April - October when resident race joined by migratory race from further south. Woodlands, mangroves, parks, urban areas.

Collared Sparrowhawk *Accipiter cirrhocephalus*
Uncommon but widespread resident, recorded in all regions. Woodlands, urban areas, parks.

Red Goshawk *Erythrotriorchis radiatus*
[3.01 3.07 3.09 4.08]
Rare but widespread resident, recorded in all regions. Open woodland, especially along and near large rivers.

Wedge-tailed Eagle *Aquila audax*
[2.02 2.03 3.07 3.10 4.01 4.05 4.06 4.08 5.02 5.03 5.04 5.05 5.08]
Moderately common and widespread, recorded in all regions. Open woodland, plains, escarpments.

Little Eagle *Hieraaetus morphnoides*
[5.02 5.05 5.07 5.08]
Uncommon visitor, perhaps resident in small numbers. Mainly southern areas, though recorded in all regions. Open woodlands, plains.

Brown Falcon *Falco berigora*
Common and widespread resident, recorded in all regions. Woodlands, floodplains, wetland edges.

Nankeen Kestrel *Falco cenchroides*
Moderately common and widespread resident, recorded in all regions. Open woodland, grasslands, parks, urban areas.

Australian Hobby *Falco longipennis*
Moderately common and widespread resident, recorded in all regions. Open woodland, wetlands, parks, urban areas.

[Oriental Hobby *Falco severus* **]**
A single claim from Darwin in June 1978 is unsubstantiated (BARC 26).

Grey Falcon *Falco hypoleucos*
[5.01 5.03 5.04 5.05 5.06 5.07 5.08]
Uncommon to rare visitor. Recorded in all regions. Open woodland, riverine woodland, wetlands, plains.

Black Falcon *Falco subniger*
[1.12 1.13 2.01 2.02 2.05 4.01 4.03 4.05 5.03 5.05]
Uncommon but widespread, recorded in all regions. May be resident in small numbers in southern areas. Most coastal and near coastal records May - October. Open woodland, wetlands, grassland.

Peregrine Falcon *Falco peregrinus*
[1.12 1.13 3.06 5.08]
Uncommon but widespread resident, recorded in all regions. Wetlands, escarpments, coasts.

Brolga *Grus rubicunda*
[1.12 1.13 2.03 2.04 3.01 3.02 3.07 3.08 5.08]
Moderately common and widespread. Recorded in all regions in all months, but greatest numbers May - November. Wetlands, grasslands.

Sarus Crane *Grus antigone*
Very rare sporadic visitor. Records from Holmes Jungle and Fogg Dam.

Buff-banded Rail *Gallirallus philippensis*
[1.05 1.09 1.10 1.12 1.14 2.01 2.03]
Moderately common resident. Recorded in all regions, though mainly coastal and near coastal. Wetlands, streams, occasionally sewage ponds, estuaries and mangroves.

Lewin's Rail *Rallus pectoralis*
Vagrant. Recorded in Casuarina Coastal Reserve (McKean 1983a).

Bush-hen *Amaurornis olivaceus*
[1.03 1.10 1.12]
Rare and localised, recorded in all regions except Southwest. Small numbers may be resident. Riverine woodland, monsoon forest, mangrove edges.

Baillon's Crake *Porzana pusilla*
[1.14 2.01]
Uncommon and irregular visitor to coastal and near coastal areas, mainly August - December. Recorded in Darwin, Fogg Dam and Kakadu regions. Wetlands.

Spotless Crake *Porzana tabuensis*
[1.14 2.03]
Rare, irregular visitor to coastal and near coastal areas. Recorded in Darwin and Fogg Dam regions. Wetlands.

White-browed Crake *Porzana cinerea*
[1.12 1.13 1.17 1.14 2.03 2.01 3.01 3.07 5.08]
Moderately common resident in coastal and near coastal areas. Recorded in all regions except Katherine. Wetlands including sewage ponds.

Chestnut Rail *Eulabeornis castaneoventris*
[1.02 1.03 1.09 1.14 1.15]
Moderately common resident in coastal mangroves. Recorded in Darwin and Kakadu regions. Mangroves.

Purple Swamphen *Porphyrio porphyrio*
[1.13 2.01 2.03 3.02 3.07 5.08]
Uncommon to moderately common and widespread visitor, mainly May - November. Recorded in all regions. Wetlands.

Black-tailed Native-Hen *Gallinula ventralis*
Rare, irregular visitor to southern Top End. Recorded in Katherine and Southwest regions. Wetlands.

Eurasian Coot *Fulica atra*

[1.10 1.14 1.17 5.08]

Uncommon to moderately common and widespread visitor, mainly May - October. Recorded in all regions, but less frequent near coast than in southern areas. Wetlands including sewage ponds.

Australian Bustard *Ardeotis australis*

[2.06 4.10 5.01 5.07 5.08]

Uncommon to rare but widespread. Recorded in all regions, though generally restricted to the southern Top End, with irregular records, mainly May - October, in more northerly areas. Open woodlands, floodplains, grasslands.

Red-backed Button-quail *Turnix maculosa*

[1.07 1.12 1.13 2.01 4.01]

Uncommon to moderately common resident. Recorded in all regions, though mainly coastal and near coastal. Wetland edges, floodplains, open woodlands.

Chestnut-backed Button-quail *Turnix castanota*

[3.05 3.06 3.07 4.01 4.02 4.04 4.06 4.07]

Uncommon and localised resident, recorded in Kakadu, Katherine and Southwest regions. Open woodlands with sparse grass cover, rocky ridges.

Red-chested Button-quail *Turnix pyrrhothorax*

[1.12 2.05 3.07]

Uncommon to moderately common resident, recorded in all regions. Open woodland, wetland edges, grasslands.

Little Button-quail *Turnix velox*

[5.04 5.08]

Rare to uncommon and localised. Recorded in Kakadu, Katherine and Southwest regions. Open woodland, grasslands.

Latham's Snipe *Gallinago hardwickii*

Vagrant. Confirmed record only from Jabiru, Kakadu region, Aug 1985 (Bywater & McKean 1987).

Pin-tailed Snipe *Gallinago stenura*

Vagrant. Single bird banded at Leanyer Swamp, Darwin region, Jan 1985 (Higgins & Davies 1996).

Swinhoe's Snipe *Gallinago megala*

[1.07 1.10 1.12 1.13 1.15 1.17 2.01 2.06 3.01 3.02 3.07 3.08]

Moderately common and widespread visitor, November - April, mainly to coastal and near coastal areas. Recorded in all regions. Wetlands, coastal grasslands, rivers, mangrove edges, occasionally beaches and sewage ponds.

Black-tailed Godwit *Limosa limosa*

[1.05 1.09 1.13 2.03 2.04]

Moderately common and widespread visitor, September - April, scarce or absent in other months. Recorded in Darwin, Fogg Dam and Kakadu regions. Beaches, mudflats, wetlands.

Bar-tailed Godwit *Limosa lapponica*

[1.01 1.05 1.06 1.08 1.09 1.14]

Common visitor to coastal areas, September - April, uncommon or absent in other months. Recorded in Darwin and Kakadu regions. Beaches, mudflats.

Little Curlew *Numenius minutus*

[1.05 1.10 1.12 1.13 1.16 1.17 2.02 2.03 3.01 3.02 3.07]

Moderately common to common and widespread passage migrant, late September - December. Rarely seen in the Top End on return migration March - April. Recorded in all regions. Grasslands, wetlands, floodplains, parks.

Whimbrel *Numenius phaeopus*

[1.01 1.05 1.06 1.07 1.09 1.14 1.15]

Common coastal visitor, September - April, uncommon in other months. Recorded in Darwin and Kakadu regions. Beaches, reefs, mudflats, mangroves.

Eastern Curlew *Numenius madagascariensis*

[1.01 1.02 1.05 1.06 1.08 1.09 1.14 1.15]

Common coastal visitor, September - April, uncommon to scarce in other months. Recorded in Darwin and Kakadu regions. Beaches, mudflats.

[Eurasian Curlew *Numenius arquata*]

A claim of this species from Darwin, Mar-Apr 1948 (Deignan 1964) is unconfirmed (Higgins & Marchant 1996).

Common Redshank *Tringa totanus*

[1.05 1.09 1.10]

Rare coastal visitor, October - April. Recorded in Darwin and Kakadu regions. Mudflats, wetlands including sewage ponds.

[Spotted Redshank *Tringa erythropus*]
A claim of this species from Darwin in 1983 (McKean & Dampney 1984) was not accepted (BARC 83).

Marsh Sandpiper *Tringa stagnatilis*
[1.10 1.12 1.13 1.17 2.01 2.03 3.02 3.07]
Common and widespread visitor, September - April, rare in other months. Recorded in all regions. Wetlands.

Common Greenshank *Tringa nebularia*
Common and widespread visitor, late August - April, uncommon in other months. Recorded in all regions. Beaches, mudflats, estuaries, wetlands including sewage ponds.

[Nordmann's Greenshank *Tringa guttifer*]
A claim of this species from Leanyer Sewage Works (McKean *et al* 1976) is unconfirmed (BARC 32).

Green Sandpiper *Tringa ochropus*
Vagrant. One accepted record, Shoal Bay Tip, March 1998 (McCrie 2000; BARC 239). A claim of this species from Kapalga, near the South Alligator River (McKean 1980a) is unconfirmed. (BARC 46)

Wood Sandpiper *Tringa glareola*
[1.11 1.13 1.17 2.01 2.03 2.06 3.07]
Moderately common and widespread visitor, late August - April, absent in other months. Recorded in all regions. Wetlands including sewage ponds, rain puddles.

Terek Sandpiper *Xenus cinereus*
[1.01 1.05 1.06 1.09 1.14 1.15]
Moderately common coastal visitor, September - April, uncommon to scarce or absent in other months. Recorded in Darwin and Kakadu regions. Beaches, mudflats, estuaries.

Common Sandpiper *Actitis hypoleucos*
Common and widespread visitor, July - April, scarce or absent in May and June. Recorded in all regions. Beaches, mudflats, estuaries, reefs, wetlands including sewage ponds.

Grey-tailed Tattler *Heteroscelus brevipes*
[1.01 1.05 1.06 1.08 1.09 1.15]
Moderately common coastal visitor, September - April, uncommon in other months. Recorded in Darwin and Kakadu regions. Reefs, estuaries, mudflats.

[Wandering Tattler *Heteroscelus incana*]
The presence of this species in the Northern Territory (Higgins & Marchant 1996) requires confirmation.

Ruddy Turnstone *Arenaria interpres*
Moderately common to common coastal visitor, September - April, generally uncommon in other months. Recorded in Darwin and Kakadu regions. Reefs, harbours, beaches.

Asian Dowitcher *Limnodromus semipalmatus*
[1.05 1.06 1.09 1.13]
Rare visitor, September - April. Recorded in Darwin region. Beaches, mudflats, wetlands.

Great Knot *Calidris tenuirostris*
[1.05 1.06 1.08 1.09]
Common coastal visitor, September - April, uncommon in other months. Recorded in Darwin and Kakadu regions. Beaches, mudflats, reefs.

Red Knot *Calidris canutus*
[1.05 1.06 1.08 1.09]
Uncommon to moderately common coastal visitor, September - April, rare to uncommon in other months. Recorded in Darwin and Kakadu regions. Beaches, mudflats.

Sanderling *Calidris alba*
[1.05 1.06 1.08 1.09]
Moderately common coastal visitor, September - April, uncommon or absent in other months. Recorded in Darwin and Kakadu regions. Beaches, mudflats and reefs.

Red-necked Stint *Calidris ruficollis*
Moderately common and widespread visitor, September - April, scarce to uncommon in other months. Recorded in Darwin, Fogg Dam and Kakadu regions. Beaches, mudflats, estuaries, wetlands including sewage ponds.

[Little Stint *Calidris minuta*]
A specimen reported to be this species was collected near Leanyer Sewage Works in November 1979 (McKean & Hertog 1981); however, there has been no formal assessment of this or subsequent claims of the species in the Northern Territory.

Long-toed Stint *Calidris subminuta*
[1.13]
Rare, irregular visitor, September - December. Recorded in Darwin region. Wetlands with short grass and muddy edges.

Baird's Sandpiper *Calidris bairdii*
Vagrant. Palmerston, October 1983 (McKean 1984; BARC 74).

Pectoral Sandpiper *Calidris melanotos*
[1.10 1.13]
Rare, irregular visitor, September - April. Recorded in Darwin and Kakadu regions.
Wetlands including sewage ponds.

Sharp-tailed Sandpiper *Calidris acuminata*
[1.13 1.17 2.01 2.03 2.06 3.07]
Moderately common and widespread visitor, late August - April, absent in other months.
Recorded in all regions. Wetlands, floodplains, sewage ponds.

Curlew Sandpiper *Calidris ferruginea*
[1.05 1.06 1.09 1.10]
Uncommon to moderately common visitor, September - April. Recorded in all regions.
Coasts, wetlands, sewage ponds.

Stilt Sandpiper *Micropalama himantopus*
Vagrant. Single record, near Leanyer Sewage Works, August-September, 1980 (McKean
et al. 1982; BARC 129).

Broad-billed Sandpiper *Limicola falcinellus*
[1.05 1.06 1.09 1.10 1.13]
Uncommon, irregular coastal visitor, September - April. Recorded in Darwin and Kakadu
regions. Coasts, wetlands, sewage ponds.

Ruff *Philomachus pugnax*
[1.11 1.13]
Rare, irregular visitor, September - April. Recorded in Darwin region. Wetlands, sewage
ponds.

Red-necked Phalarope *Phalaropus lobatus*
[1.10 1.14]
Rare, irregular visitor, November - March. Recorded in Darwin region. Coasts, wetlands, sewage ponds.

Comb-crested Jacana *Irediparra gallinacea*
Common and widespread resident, recorded in all regions. Wetlands including sewage
ponds.

Bush Stone-curlew *Burhinus grallarius*
[1.05 1.12 1.13 1.16 1.17 2.01 2.06 3.05 3.07 3.10 4.01 5.06 5.08]
Moderately common and widespread resident. Recorded in all regions. Woodlands, orchards, parks, occasionally urban areas.

Beach Stone-curlew *Esacus neglectus*
[1.01 1.05 1.06 1.08 1.09 1.14]
Uncommon resident along coasts. Recorded in Darwin and Kakadu regions. Beaches, reefs, mudflats.

Pied Oystercatcher *Haematopus longirostris*
[1.05 1.06 1.08 1.09]
Uncommon to moderately common coastal resident. Recorded in Darwin and Kakadu regions. Beaches, reefs, mudflats.

Sooty Oystercatcher *Haematopus fuliginosus*
[1.05 1.06 1.08 1.09]
Uncommon coastal resident, with some movement between mainland and offshore islands. Most frequently recorded in May - October. Recorded in Darwin and Kakadu regions. Beaches, reefs.

Black-winged Stilt *Himantopus himantopus*
Common and widespread resident. Recorded in all regions. Wetlands, sewage ponds.

Red-necked Avocet *Recurvirostra novaehollandiae*
[1.10 1.12 2.03 2.06]
Rare, irregular visitor, with scattered records, mainly May - October. Recorded in all regions. Wetlands, sewage ponds.

Pacific Golden Plover *Pluvialis fulva*
[1.05 1.06 1.08 1.09 1.13]
Moderately common and widespread visitor, September - May, uncommon in other months. Recorded in Darwin, Fogg Dam and Kakadu regions. Beaches, mudflats, reefs, wetlands, grasslands, sewage ponds.

Grey Plover *Pluvialis squatarola*
[1.05 1.06 1.07 1.08 1.09]
Moderately common coastal visitor, September - April, uncommon in other months. Recorded in Darwin and Kakadu regions. Beaches, mudflats, reefs.

Ringed Plover *Charadrius hiaticula*
Vagrant. A report of this species from Leanyer Sewage Works in February 1980 (McKean 1983) is very credible, but has not been assessed by BARC.

Little Ringed Plover *Charadrius dubius*
[1.10 1.11 1.13]
Uncommon to rare visitor, August - April. Recorded in Darwin and Kakadu regions. Wetlands, sewage ponds.

Kentish Plover *Charadrius alexandrinus*
Vagrant. Single record, Buffalo Creek, November 1988 (McCrie 1995; BARC 170)

Red-capped Plover *Charadrius ruficapillus*
[1.05 1.06 1.07 1.08 1.09 1.10 1.14]
Moderately common resident, mainly in coastal areas. Recorded in all regions. Beaches, wetlands, sewage ponds.

Lesser Sand Plover *Charadrius mongolus*
[1.01 1.05 1.06 1.07 1.08 1.09 1.14 1.15]
Moderately common coastal visitor, September - April, less common in other months. Recorded in Darwin and Kakadu regions. Beaches, mudflats.

Greater Sand Plover *Charadrius leschenaultii*
[1.01 1.05 1.06 1.07 1.08 1.09 1.14 1.15]
Common coastal visitor, September - April, with significant numbers overwintering. Recorded in Darwin and Kakadu regions. Beaches, mudflats.

Caspian Plover *Charadrius asiaticus*
Vagrant. There is a specimen from Pine Creek, September 1896 (Condon 1961) and a sight-record from Lake Finniss, October 1994 (McCrie & Jaensch 1999; BARC 218). A report of this species from Leanyer Swamp (McKean *et al.* 1976) is unsubstantiated.

Oriental Plover *Charadrius veredus*
[1.05 1.06 1.10 1.12 1.13 2.02 2.03]
Uncommon to moderately common passage visitor, August - December, rare or absent during northward migration. Recorded in all regions. Mudflats, reefs, wetlands, flood-plains, grasslands, sewage ponds.

Red-kneed Dotterel *Erythrogonys cinctus*
[1.13 2.01 2.03 3.02 4.01]
Uncommon to moderately common and widespread. Most common April - October though present all year in small numbers. Recorded in all regions. Wetlands, sewage ponds.

Black-fronted Dotterel *Elseyornis melanops*
[1.11 1.13 1.17 2.01 2.03 2.06 3.02 3.07 4.01]
Moderately common and widespread resident, with movement away from coastal areas December - March. Recorded in all regions. Wetlands, sewage ponds, streams.

Masked Lapwing *Vanellus miles*
Very common and widespread resident. Recorded in all regions. Mudflats, reefs, wetlands, parks, urban areas.

Oriental Pratincole *Glareola maldivarum*
[1.10 1.12 1.13 1.17 2.01 3.02]
Uncommon to moderately common and widespread passage visitor, October - December. Absent or rare on northward passage. Recorded in all regions. Wetlands, floodplains, grasslands, sewage ponds.

Australian Pratincole *Stiltia isabella*
[1.10 1.13 1.17 2.01 2.02 3.01 3.02 3.06 3.07 3.08 4.01 4.08 5.05 5.08]
Common and widespread visitor, April - November, with occasional records other months including breeding. Recorded in all regions. Wetlands, grasslands, plains, sewage ponds.

Arctic Jaeger *Stercorarius parasiticus*
[1.08]
Rare, irregular visitor to coast. Recorded in Darwin region. Inshore, harbours.

Pomarine Jaeger *Stercorarius pomarinus*
[1.01]
Rare, irregular visitor to coast. Recorded in Darwin region. Inshore, harbours.

Black-tailed Gull *Larus crassirostris*
Vagrant. Single record, Darwin, April - September 1982 (McKean & Thompson 1983).

Silver Gull *Larus novaehollandiae*
[1.01 1.06 1.08 1.09]
Common only along Darwin coastal area. Recorded in all regions except Fogg Dam. Beaches, reefs, harbours, rubbish tips, sewage ponds.

Black-headed Gull *Larus ridibundus*
Vagrant. Records from Buffalo Creek, Lee Point and Darwin Harbour, Feb - March 1998, Dec 2002 - Apr 2003 (McCrie & McCrie 1999; BARC 237)

Sabine's Gull *Larus sabini*
Vagrant. Single record, Darwin Harbour, April 1982 (Shannon & McKean 1983; BARC 86)

Gull-billed Tern *Sterna nilotica*
[1.05 1.09 1.10 1.13 2.01 2.03 3.01 3.02 3.07]
Moderately common and widespread. Recorded in Darwin, Fogg Dam and Kakadu regions. Inshore, beaches, reefs, harbours, wetlands, sewage ponds.

Caspian Tern *Sterna caspia*
Moderately common and widespread resident. Recorded in all regions. Mudflats, beaches, wetlands.

Crested Tern *Sterna bergii*
Common coastal resident (breeding on offshore islands), recorded in Darwin and Kakadu regions. Inshore, beaches, reefs, harbours.

Lesser Crested Tern *Sterna bengalensis*
[1.01 1.05 1.06 1.08 1.09]
Moderately common, coastal, with some breeding. Recorded in Darwin and Kakadu regions. Greatest numbers July - November. Inshore, reefs, beaches.

Roseate Tern *Sterna dougallii*
[1.08 1.09]
Rare except in north-east Top End, coastal. Irregular visitor away from north-east coast, mainly November - April. Recorded in Darwin and Kakadu regions. Inshore, beaches.

Black-naped Tern *Sterna sumatrana*
Rare except in north-east Top End, coastal. Irregular visitor away from north-east coast. Recorded in Darwin and Kakadu regions. Inshore, beaches.

Common Tern *Sterna hirundo*
[1.01 1.05 1.08 1.09 1.10]
Uncommon to moderately common coastal visitor, October - April, scarce or absent in other months. Recorded in Darwin and Kakadu regions. Beaches, harbours, sewage ponds.

Little Tern *Sterna albifrons*
[1.01 1.05 1.06 1.08 1.09]
Uncommon to moderately common along coast, with small numbers breeding mainly in north-east. Recorded in Darwin and Kakadu regions. Most common September - April. Beaches, harbours.

Bridled Tern *Sterna anaethetus*
[1.01 1.08]
Rare except in north-east Top End, coastal. Irregular visitor away from north-east coast. Recorded in Darwin and Kakadu regions. Inshore, harbours.

Sooty Tern *Sterna fuscata*
[1.08 1.09]
Rare except in north-east Top End, coastal. Recorded in Darwin region. Inshore, beaches.

Whiskered Tern *Chlidonias hybridus*
Common and widespread visitor, mainly March - November, uncommon to scarce in other months. Recorded in all regions. Inshore, harbours, wetlands, floodplains, sewage ponds.

White-winged Black Tern *Chlidonias leucopterus*
[1.10 1.11 1.13 1.14 1.17 3.02 3.07]
Common coastal and near coastal visitor, September - April, scarce or absent in other months. Recorded in Darwin, Fogg Dam and Kakadu regions. Inshore, beaches, wetlands, sewage ponds.

Common Noddy *Anous stolidus*
Rare except in north-east Top End, coastal. Recorded in Darwin region. Inshore, beaches, sewage ponds.

Black Noddy *Anous minutus*
Vagrant offshore (Higgins & Davies 1996).

Rock Dove *Columba livia*
Common urban resident. Introduced. Recorded in Darwin, Katherine and Southwest regions.

Emerald Dove *Chalcophaps indica*
[1.05 1.07 1.09 1.12 1.16 2.01 3.05 3.06]
Moderately common resident, recorded in all regions. Monsoon forest, occasionally mangroves, parks.

Common Bronzewing *Phaps chalcoptera*
[2.05 3.07 3.08 3.09 3.10 4.01 4.08 5.02 5.04 5.06 5.08]
Moderately common resident, widespread away from coast. Recorded in all regions except Fogg Dam. Open woodland.

Flock Bronzewing *Phaps histrionica*
[1.08 1.10]
Rare, irregular visitor, April - October. Scattered records throughout much of Top End. Recorded in all regions except Fogg Dam. Floodplains, wetland edges, sewage ponds.

Crested Pigeon *Ocyphaps lophotes*
[4.01 4.06 4.07 4.08 4.09 4.10 5.02 5.03 5.05 5.08]
Moderately common and widespread resident. Recorded in all regions, though rarely nearer coast. Open woodland, urban areas of southern Top End.

Spinifex Pigeon *Geophaps plumifera*
[5.04 5.05 5.06 5.08]
Uncommon to moderately common resident in southern Top End. Recorded in Kakadu, Katherine and Southwest regions. Open woodland, ridges, spinifex-covered or grassy hillsides, particularly in Southwest.

Partridge Pigeon *Geophaps smithii*
[3.05 3.07 3.09 3.10 3.11 4.01 4.02]
Uncommon to moderately common resident. Recorded in all regions except Southwest. Open woodland, particularly in Kakadu and surrounding areas.

White-quilled Rock-Pigeon *Petrophassa albipennis*
[5.03 5.08]
Uncommon and localised resident in Southwest escarpment areas. Recorded in Southwest region only. Sandstone uplands and escarpments.

Chestnut-quilled Rock-Pigeon *Petrophassa rufipennis*
[3.04 3.05 3.06 3.09 3.10]
Uncommon to moderately common and localised resident. Recorded in Kakadu region only. Sandstone uplands and escarpments of Kakadu National Park and surrounding areas.

Diamond Dove *Geopelia cuneata*
[1.17 2.05 3.11 4.01 4.07 4.08 4.10 5.01 5.02 5.04 5.06 5.08]
Common, widespread away from coast. All year, with some birds moving into near coastal areas May - October.

Peaceful Dove *Geopelia placida*
Common and widespread resident. Recorded in all regions. Woodland, wetland edges, parks, urban areas.

Bar-shouldered Dove *Geopelia humeralis*
Common and widespread resident, recorded in all regions. Woodland, wetland edges, parks, urban areas.

Banded Fruit-Dove *Ptilinopus alligator*
[3.05 3.06 3.09 3.10 3.11]
Uncommon and localised resident in and around Kakadu. Sandstone uplands with monsoon forest or rock figs.

Rose-crowned Fruit-Dove *Ptilinopus regina*
[1.05 1.07 1.09 1.12 1.16 2.01 2.06 3.03 3.10 3.11]
Moderately common resident. Recorded in all regions, though mainly coastal and near coastal. Monsoon forest and adjacent woodland, mangrove edges, urban areas.

Elegant Imperial-Pigeon *Ducula concinna*
Vagrant. Single record, Nightcliff, 1994 (Wright et al. 1995; BARC 181).

Torresian Imperial-Pigeon *Ducula spilorrhoa*
Common and widespread breeding visitor, July - April, with relatively small but increasing numbers present throughout the year, particularly in some Darwin suburban areas. Recorded in all regions. Monsoon forests, riverine woodland, parks and other urban environments.

Red-tailed Black-Cockatoo *Calyptorhynchus banksii*
Common and widespread resident, though more common coastally May - September. Recorded in all regions. Open woodland, wetland edges, urban areas. Often seen feeding on ground in burnt out areas of woodland.

Galah *Cacatua roseicapilla*
Common and widespread resident, recorded in all regions. Woodland, wetland edges, parks, urban areas.

Sulphur-crested Cockatoo *Cacatua galerita*
Common and widespread resident. Recorded in all regions. Woodland, wetland edges, riverine forest, parks and gardens.

Little Corella *Cacatua sanguinea*
Common and widespread resident. Recorded in all regions. Wetland edges, woodlands, parks.

Cockatiel *Nymphicus hollandicus*
[4.01 4.03 4.04 4.06 4.08 4.09 4.10 5.01 5.02 5.03 5.05 5.08]
Moderately common resident, widespread away from coast. Recorded in all regions but only common in Katherine and Southwest regions. Occasional influxes to coastal areas May - October in dry years. Open woodland.

Rainbow Lorikeet *Trichoglossus haematodus*
Common and widespread resident, recorded in all regions. Woodland, parks, urban areas.

Varied Lorikeet *Psitteuteles versicolor*
[1.08 1.12 1.14 1.16 1.17 2.06 3.05 3.09 3.11 4.06 4.08 5.05 5.06]
Moderately common, and widespread resident. Recorded in all regions, though more common away from coast. Highly nomadic, moving into and out of areas following flowering trees, so can be difficult to anticipate. Check areas where eucalypts are in flower. Open woodland, riverine woodland.

Red-winged Parrot *Aprosmictus erythropterus*
Common and widespread resident, recorded in all regions. Woodlands, parks, urban areas.

Northern Rosella *Platycercus venustus*
[1.03 1.14 1.15 1.17 1.18 2.05 2.06 3.03 3.05 3.06 4.01 4.02 4.03 5.02]
Uncommon to moderately common and widespread resident, but patchy. Recorded in all regions, though least common in Darwin and Fogg Dam regions. Woodland, riverine forest.

Hooded Parrot *Psephotus dissimilis*
[3.11 4.01 4.02 4.03 4.04 4.06 4.07 4.08 4.09 4.10]
Uncommon, localised in Katherine and Kakadu regions. All year. Open woodland, riverine woodland.

Budgerigar *Melopsittacus undulatus*
[5.01 5.04 5.08]
Rare to uncommon irregular visitor, not recorded in Darwin and Fogg Dam regions. Most records from southern Top End, particularly Southwest region, May - October. Open woodland.

[Night Parrot *Pezoporus occidentalis*]
Unconfirmed report, Keep River, April 1982 (McKean 1985).

Oriental Cuckoo *Cuculus saturatus*
[1.05 1.07 1.08 1.09 1.12 1.17 2.01 2.02 2.04 2.06 3.01 3.02 3.07 3.08 3.11 4.08]
Uncommon but relatively widespread visitor, November - April. Recorded in all regions. Monsoon forest, riverine woodlands, wetland edges (particularly paperbark swamps), orchards, occasionally urban areas. Often at edges of dense vegetation, where it can feed on relatively open ground but readily return to cover.

Pallid Cuckoo *Cuculus pallidus*
[1.05 1.08 1.13 3.06 4.01 4.03 5.04 5.05 5.07 5.08]
Uncommon to moderately common and widespread away from coast. Recorded in all regions, but more common in Southwest region. Records for all months but most common May - November, when there is also some movement towards the coast. Woodland, wetland edges.

Brush Cuckoo *Cacomantis variolosus*
[1.04 1.05 1.13 1.17 2.01 2.02 2.03 3.01 3.02 3.05 3.07 4.03 4.08]
Moderately common and widespread resident. Recorded in all regions. Most easily seen September - April, when it is more vocal than at other times of the year. Monsoon forest edges, woodlands, wetland edges, parks, occasionally streets and gardens.

Little Bronze-Cuckoo *Chrysococcyx minutillus*
[1.03 1.04 1.05 1.07 1.09 1.15 2.01 2.04 3.03 3.06 3.07 3.08 3.11]
Moderately common and widespread resident, recorded in all regions but more common nearer coast. Mangroves, monsoon forest, wetland edges, riverine forest, occasionally parks and gardens.

Horsfield's Bronze-Cuckoo *Chrysococcyx basalis*
[2.05 5.01 5.04 5.05]
Uncommon to moderately common and widespread visitor. Recorded in all regions, but mostly in the southern and central Top End. Records in all months but mainly June - October, when a few birds also move to more coastal areas.

Black-eared Cuckoo *Chrysococcyx osculans*
[2.01 3.02 3.07 5.03]
Rare, irregular visitor, with records throughout the year. Recorded in all regions. Most reports are from southern areas, with sporadic northerly records May - October.

Australian Koel *Eudynamys cyanocephala*
[1.07 1.09 2.01 2.04 2.06 3.01 3.02 3.05 3.06 3.07 3.10]
Moderately common and widespread breeding visitor, September - April. Recorded in all regions. Woodlands, monsoon forest, wetland edges, parks, urban areas.

[Long-tailed Cuckoo *Eudynamis taitensis*]
A claim of this species from Darwin in May 1980 is unsubstantiated (BARC 39).

Channel-billed Cuckoo *Scythrops novaehollandiae*
[1.05 1.09 2.01 4.03]
Uncommon but relatively widespread breeding visitor, mainly September - April, with a few records in other months. Recorded in all regions. Woodlands, wetland edges, parks, urban areas.

Pheasant Coucal *Centropus phasianinus*
Common and widespread resident, recorded in all regions. Woodlands, riverine woodland, wetland edges, urban areas.

Rufous Owl *Ninox rufa*
[1.04 1.07 1.12 1.16 3.01 3.03 3.10 4.08]
Uncommon, mainly coastal and near coastal resident. Recorded in all regions except Southwest. Monsoon, paperbark and riverine forests, parks.

Barking Owl *Ninox connivens*
[1.04 1.05 2.01 2.03 3.01 3.02 3.06 3.07 3.08 3.10 3.11 4.08 5.03 5.05 5.08]
Moderately common and widespread resident, recorded in all regions. Woodland, monsoon forest, wetland edges, parks, urban areas.

Southern Boobook *Ninox boobook*
[2.05 3.08 5.03 5.04 5.05]
Uncommon to moderately common and widespread resident. Recorded in all regions. Open woodlands, riverine woodland, parks, urban areas.

Masked Owl *Tyto novaehollandiae*
[2.02]
Rare resident in coastal and near coastal areas, recorded in Fogg Dam and Kakadu regions. Open woodland.

Barn Owl *Tyto alba*
[1.08 2.02 3.01]
Uncommon but widespread resident, recorded in all regions. Woodlands, escarpments, wetland edges.

Eastern Grass Owl *Tyto longimembris*
[1.12 3.01]
Rare. Status and distribution uncertain, but recorded in all regions. At least some likely to be resident. Wetland edges, floodplains, grasslands.

Tawny Frogmouth *Podargus strigoides*
[1.07 1.09 2.01 3.01 3.08 5.05 5.08]
Moderately common and widespread resident, recorded in all regions. Open woodland, monsoon forest, wetland edges, parks, urban areas.

Large-tailed Nightjar *Caprimulgus macrurus*
[1.05 1.07 1.09 1.12 1.16 3.01]
Moderately common coastal and near coastal resident. Recorded in Darwin, Fogg Dam and Kakadu regions. Mangroves, monsoon and riverine forests.

Spotted Nightjar *Eurostopodus argus*
[1.13 1.14 2.02 3.07 3.08 4.06 4.08 5.05]
Moderately common and widespread resident. Recorded in all regions but largely absent from coastal areas December - April. Woodlands, wetland edges, parks, occasionally urban areas. Most frequently seen along roads at night .

Australian Owlet-nightjar *Aegotheles cristatus*
[1.05 1.09 1.12 3.01 3.07 3.08 4.07 4.08]
Moderately common and widespread resident, recorded in all regions. Woodlands. Often sits on roads at night.

Fork-tailed Swift *Apus pacificus*
Moderately common and widespread visitor, October - April. Recorded in all regions. Aerial, over all habitats.

House Swift *Apus nipalensis*
Vagrant. Single record from Point Stuart, Fogg Dam region, March 1979 (Robertson 1980; BARC 91)

Azure Kingfisher *Alcedo azurea*
[1.01 1.09 1.10 1.12 1.14 1.16 2.01 2.04 3.06 3.07 3.08 4.03]
Common and widespread resident, recorded in all regions. Streams and pools in monsoon forest and mangroves, wetlands, rivers and creeks.

Little Kingfisher *Alcedo pusilla*
[1.02 1.07 1.09 1.14 1.16 2.01 2.06 3.07 3.08]
Uncommon to moderately common resident. Recorded in all regions except Southwest. Streams and pools in monsoon forest and mangroves, densely vegetated wetlands, rivers and creeks.

Blue-winged Kookaburra *Dacelo leachii*
Common and widespread resident, recorded in all regions. Woodlands, wetland edges, parks, occasionally urban areas.

Forest Kingfisher *Todiramphus macleayii*
[1.03 1.04 1.07 1.12 1.16 2.01 2.02 2.03 2.04]
Common and widespread resident, recorded in all regions, though more common in northern areas. Riverine woodland (especially paperbark forests), wetland edges, parks, streets and gardens.

Red-backed Kingfisher *Todiramphus pyrrhopygia*
[1.05 1.14 1.15 2.02 2.04 4.03 5.07]
Moderately common and widespread. Some resident, with an influx from further inland May - October, when there is also some movement to more coastal areas. Recorded in all regions. Wetland edges, woodlands, parks. Like other species of woodland kingfishers (Sacred, Forest, Blue-winged Kookaburra), often perched on overhead power lines by roadsides.

Collared Kingfisher *Todiramphus chloris*
[1.02 1.05 1.06 1.09 1.15]
Moderately common resident in coastal areas. Recorded in Darwin and Kakadu regions. Mangroves, reefs. Confusion is possible with Sacred Kingfisher, which also occurs in these habitats.

Sacred Kingfisher *Todiramphus sanctus*
Common and widespread. Recorded in all regions. There is an influx of birds from March, with some continuing on passage, others remaining until about October. Woodlands, wetlands, mangroves, reefs, urban areas.

Rainbow Bee-eater *Merops ornatus*
Common and widespread, recorded in all regions. Some birds are resident, but the majority are visitors from April - October. Woodlands, wetland edges, coastal scrubs, urban areas.

Dollarbird *Eurystomus orientalis*
Common and widespread breeding visitor, August - May. Recorded in all regions. Open woodlands, wetlands, riverine woodland, urban areas.

Rainbow Pitta *Pitta iris*
[1.05 1.09 1.12 1.15 1.16 2.01 3.01 3.03 3.06 3.10]
Moderately common resident. Recorded in all regions except Southwest, though more common nearer the coast. Monsoon forest, occasionally forests adjoining mangroves. Easiest to observe September - March, when it is most vocal and active.

Black-tailed Treecreeper *Climacteris melanura*
[2.05 2.06 3.05 3.06 3.08 3.10 4.02 4.03 4.06 4.07 5.04 5.06]
Moderately common and widespread resident. Recorded in all regions, though more common away from the coast. Open woodland.

Red-backed Fairy-wren *Malurus melanocephalus*
Common and widespread resident, recorded in all regions. Woodlands, grasslands. The only fairy-wren in much of the Top End.

Variegated Fairy-wren *Malurus lamberti*
[3.05 3.10 5.02 5.03 5.08]
Rare to moderately common resident, recorded in all regions except Darwin and Fogg Dam. Savanna woodland, escarpment, rocky hills, spinifex woodland. Race *M. l. dulcis* (Lavender-flanked Fairy-wren) occurs on sandstone escarpments in Kakadu and Katherine regions.

Purple-crowned Fairy-wren *Malurus coronatus*
[5.03 5.05 5.06]
Uncommon and localised resident in Southwest region. Inhabits dense grasses along Victoria River and some of its offshoots. Its distinctive call, atypical for a fairy-wren, often makes it easy to locate.

White-throated Grasswren *Amytornis woodwardi*
[3.09 3.10]
Uncommon and localised resident. Recorded in Kakadu and Katherine regions. Sandstone uplands and escarpments with spinifex.

Red-browed Pardalote *Pardalotus rubricatus*
[4.09 5.01 5.02 5.04 5.05 5.07]
Uncommon to moderately common resident in southern areas. Recorded in all regions except Fogg Dam. There may be some northward movement May - October. Open woodlands near water, along dry and near dry rivers and creeks. The call is distinctive, but ventriloquial, and even when heard quite near the bird can be difficult to locate.

Striated Pardalote *Pardalotus striatus*
Common and widespread resident. Recorded in all regions. There is an influx to more coastal areas March - October, during which time they breed and are particularly vocal. Roadsides, open woodlands, creeks, rivers, parks.

Weebill *Smicrornis brevirostris*
Common and widespread resident, recorded in all regions. Woodlands.

Green-backed Gerygone *Gerygone chloronotus*
[1.04 1.05 1.07 1.09 1.12 1.15 1.16 2.01 2.04 3.05 3.08]
Moderately common and widespread resident. Monsoon forest, landward edge of mangroves, riverine forest, parks, gardens adjacent to coastal bush. Distinctive reeling call carries quite some distance, and birds can be readily located.

White-throated Gerygone *Gerygone olivacea*
[2.05 4.03 4.06 4.07 4.08 4.09]
Uncommon to moderately common and widespread, generally away from the coast. Some resident, with presumably some movement into the Top End, as well as toward more coastal areas, May - October. Woodlands, particularly in southern Top End. Its distinctive silvery call carries some distance and makes bird quite easy to locate.

Large-billed Gerygone *Gerygone magnirostris*
[1.02 1.03 1.05 1.07 1.09 1.15 2.04]
Moderately common and widespread resident. Recorded in all regions, though more common nearer the coast. Mangroves, monsoon forest, dense riverine forest.

Mangrove Gerygone *Gerygone levigaster*
[1.02 1.03 1.05 1.07 1.09 1.15]
Moderately common and widespread resident in coastal mangroves. Recorded in Darwin and Kakadu regions. Mangroves, particularly edges adjoining saline flats.

Dusky Honeyeater *Myzomela obscura*
Common and widespread resident, recorded in all regions. Forests, wetland edges, parks, urban areas.

Red-headed Honeyeater *Myzomela erythrocephala*
[1.02 1.03 1.05 1.07 1.09 1.14 1.15 2.04]
Common resident in coastal and near coastal areas. Recorded in all regions except Katherine. Inhabits mangroves, moving into nearby parks and gardens when there is little flowering in mangroves.

Banded Honeyeater *Certhionyx pectoralis*
[2.05 3.05 3.06 3.09 3.10 3.11 4.07 4.08 4.09 5.01 5.02 5.04 5.06 5.08]
Common resident, widespread away from coast, with irregular movement into more coastal areas May - October. Recorded in all regions. Open woodlands, escarpments.

Black Honeyeater *Certhionyx niger*
Rare visitor or vagrant to southern Top End. Recorded in Katherine and Southwest regions.

Brown Honeyeater *Lichmera indistincta*
Very common and widespread resident, recorded in all regions. Occurs in almost all habitats.

White-lined Honeyeater *Meliphaga albilineata*
[3.05 3.06 3.09 3.10]
Uncommon and localised resident. Recorded in sandstone escarpment and surrounding woodlands in Kakadu and Southwest regions.

Singing Honeyeater *Lichenostomus virescens*
[2.05 4.08 5.03 5.04 5.05 5.08]
Moderately common and widespread resident, recorded in all regions. Generally scarce near the coast to where there is irregular movement, May - October. Open woodland.

White-gaped Honeyeater *Lichenostomus unicolor*
Common and widespread resident, recorded in all regions. Woodlands, forests, wetland edges, riverine woodland, parks, urban areas.

Grey-headed Honeyeater *Lichenostomus keartlandi*
[5.04]
Rare in the southern Top End, nomadic. Recorded in Southwest region. Woodland, particularly on hillsides with spinifex.

Yellow-tinted Honeyeater *Lichenostomus flavescens*
[4.01 4.02 4.03 4.06 4.08 4.09 4.10 5.03 5.05 5.06]
Moderately common resident, widespread away from coast. All regions except Darwin and Fogg Dam. Open or riverine woodland, parks, urban areas.

Grey-fronted Honeyeater *Lichenostomus plumulus*
[4.08 4.09 4.10 5.01 5.04 5.05 5.06 5.07 5.08]
Uncommon, nomadic, with perhaps small numbers resident. Recorded in southern areas of Katherine and Southwest regions only. Open woodlands.

White-throated Honeyeater *Melithreptus albogularis*
Common and widespread resident. Recorded in all regions. Eucalypt forests and woodlands, wetland edges, parks, urban areas.

Black-chinned Honeyeater *Melithreptus laetior*
[4.01 4.06 4.07 4.09 5.04 5.06]
Uncommon resident, widespread across southern Top End. Recorded in all regions except Darwin and Fogg Dam. Woodland, wetland edges, creeks and rivers with or without water.

Little Friarbird *Philemon citreogularis*
Common and widespread resident. Recorded in all regions. Woodland, wetland edges, mangrove edges, parks, urban areas.

Helmeted Friarbird *Philemon buceroides*
[1.03 1.05 1.07 1.09 3.05 3.06 3.10]
Moderately common resident. Recorded in all regions except Fogg Dam and Southwest. Mangroves, monsoon forest, escarpment, parks, streets, gardens.

Silver-crowned Friarbird *Philemon argenticeps*
Common and widespread resident. Recorded in all regions. Woodland, escarpment, parks, urban areas.

Bar-breasted Honeyeater *Ramsayornis fasciatus*
[1.16 2.01 3.02 3.06 3.07 3.10 3.11 4.04 4.08]
Moderately common and widespread resident. Recorded in all regions. Wetland edges, parks, urban areas.

Rufous-banded Honeyeater *Conopophila albogularis*
Common and widespread resident, though mainly coastal and near coastal. Recorded in all regions. Monsoon forest, mangroves, riverine woodlands, wetland edges, parks, urban areas.

Rufous-throated Honeyeater *Conopophila rufogularis*
[2.05 3.05 3.09 3.10 3.11 4.03 4.07 4.09 5.01 5.04 5.08]
Common and widespread resident, recorded in all regions. More common away from coast, but with some movement toward the coast May - October. Riverine and open woodlands, wetland edges, escarpments.

136

Blue-faced Honeyeater *Entomyzon cyanotis*
Moderately common and widespread resident. Recorded in all regions. Woodlands, parks, urban areas.

Yellow-throated Miner *Manorina flavigula*
[1.14 1.17 2.05 2.06 4.01 4.02 4.08 5.01 5.04 5.07 5.08]
Moderately common and widespread resident. Recorded in all regions, though generally away from coast. Woodlands, parks, urban areas.

Crimson Chat *Epthianura tricolor*
[5.01 5.04 5.05]
Rare and irregular visitor, mainly May - October. Recorded in Southwest region only. Woodlands.

Yellow Chat *Epthianura crocea*
[3.01]
Rare and localised resident and nomadic visitor. Recorded in all regions. Wetland edges.

Jacky Winter *Microeca fascinans*
[2.05 4.03 4.04 4.08 4.10 5.02 5.06 5.08]
Moderately common resident away from coast. Recorded in all regions except Darwin. Riverine and open woodland, floodplain edges.

Lemon-bellied Flycatcher *Microeca flavigaster*
[1.02 1.05 1.07 1.09 2.01 2.04 2.05 3.02 3.07 3.08]
Common and widespread resident. Recorded in all regions. Monsoon forest, mangroves, riverine woodlands, parks.

Red-capped Robin *Petroica goodenovii*
[5.04]
Rare, irregular visitor to Southwestern Top End, May - October. Recorded in Southwest region only. Woodlands.

Hooded Robin *Melanodryas cucullata*
[3.11 4.03 4.10 5.04]
Uncommon to moderately common resident, widespread in southern Top End, localised in more northerly areas. Recorded in all regions except Darwin and Fogg Dam. Woodlands.

Mangrove Robin *Eopsaltria pulverulenta*
[1.02 1.03 1.14 1.15]
Moderately common coastal resident. Recorded in Darwin and Kakadu regions. Mangroves.

White-browed Robin *Poecilodryas superciliosa*
[2.05 3.02 3.08 3.10 3.11 4.08 5.05]
Uncommon resident, widespread in suitable habitat. Recorded in all regions except Darwin. Dense riverine vegetation.

Grey-crowned Babbler *Pomatostomus temporalis*
Common and widespread resident. Recorded in all regions. Open woodlands, occasionally parks.

Varied Sittella *Daphoenositta chrysoptera*
[2.05 4.01 4.03 4.06 4.07 4.08 4.09 4.10 5.04]
Uncommon to moderately common resident and nomad, widespread away from coast. Recorded in all regions. Woodlands.

Crested Shrike-tit *Falcunculus frontatus*
[4.07 4.09 4.10]
Rare and little known nomadic resident, localised in southern and eastern Top End. Recorded in Kakadu and Katherine regions. Open woodlands, occasionally wetland edges.

Grey Whistler *Pachycephala simplex*
[1.05 1.09 1.16 2.01 2.04]
Common and widespread resident, recorded in all regions. Mangroves, monsoon forest.

Mangrove Golden Whistler *Pachycephala melanura*
[1.02 1.09 1.14 2.04]
Uncommon resident. Recorded in all regions except Katherine and Southwest. Mangroves, mainly coastal but occasionally estuarine, adjacent vine thickets.

Rufous Whistler *Pachycephala rufiventris*
Common and widespread resident. Recorded in all regions. Some movement to more coastal areas May to October. Woodlands.

White-breasted Whistler *Pachycephala lanioides*
[1.06 1.14]
Rare coastal resident. Recorded in Darwin and Kakadu regions. Mangroves.

Little Shrike-thrush *Colluricincla megarhyncha*
[1.05 1.09 1.16 2.01 3.01 3.03 3.04 3.07 3.08 3.10]
Uncommon to moderately common resident, widespread. Recorded in all regions. Mangroves, monsoon forest and dense riverine vegetation.

Sandstone Shrike-thrush *Colluricincla woodwardi*
[3.03 3.05 3.06 3.10 5.03 5.08]
Uncommon localised resident. Recorded in all regions except Darwin and Fogg Dam. Restricted to sandstone escarpments.

Grey Shrike-thrush *Colluricincla harmonica*
Moderately common to common resident, widespread except in coastal areas. Recorded in all regions. Open woodland.

Willie Wagtail *Rhipidura leucophrys*
Common and widespread resident. Recorded in all regions, though generally absent from northern coastal areas October - April. Woodlands, edges of wetlands, parks, urban areas.

Northern Fantail *Rhipidura rufiventris*
Common and widespread resident, recorded in all regions. Open and riverine woodlands, mangroves, parks, urban areas.

Mangrove Grey Fantail *Rhipidura phasiana*
[1.09 1.14 1.15]
Uncommon coastal resident. Recorded in Darwin and Kakadu regions. Mangroves.

Grey Fantail *Rhipidura fuliginosa*
Uncommon to rare, probably migratory. Status uncertain, with some claims presumably referring to Northern Fantail. Mainly in the south, with occasional influxes further north from May - October. Open and riverine woodland, occasionally mangroves.

Rufous Fantail *Rhipidura rufifrons*
[1.07 1.09 1.14 2.04 3.07 3.08]
Uncommon to moderately common resident. Recorded in all regions. Monsoon forest, mangroves, dense riverine vegetation.

Spangled Drongo *Dicrurus bracteatus*
Common and widespread resident, with local movements. Recorded in all regions. Monsoon and riverine forest, wetland edges, parks, urban areas.

Black-faced Monarch *Monarcha melanopsis*
Vagrant. Single record, Howard Springs.

Spectacled Monarch *Monarcha trivirgatus*
Vagrant. Single record, Groote Eylandt (Noske & Brennan 2002)

Leaden Flycatcher *Myiagra rubecula*
[1.09 1.16 2.01 2.03 2.04 3.01 3.02 3.08 3.10 4.03 4.06]
Common and widespread resident, with local movements. Recorded in all regions. Open and riverine woodlands, mangroves, monsoon forest, wetland edges, parks and urban areas.

Broad-billed Flycatcher *Myiagra ruficollis*
[1.07 1.09 1.10 1.14 2.01 2.04 3.07]
Moderately common resident of coastal and near coastal areas. Recorded in all regions except Southwest. Mangroves, monsoon forest, paperbark woodland.

Restless Flycatcher *Myiagra inquieta*
[1.13 1.17 2.01 2.02 2.03 3.01 3.02 3.07 3.08 4.03 4.08 4.10]
Moderately common and widespread resident. Recorded in all regions. Paperbark and riverine forest, edges of wetlands, floodplains.

Shining Flycatcher *Myiagra alecto*
[1.12 1.18 2.01 2.04 3.04 3.05 3.07 3.08 3.10 3.11 4.08 5.03]
Common and widespread resident. Recorded in all regions. Mangroves, monsoon and riverine forests.

Magpie-lark *Grallina cyanoleuca*
Common and widespread. Recorded in all regions. Some birds resident, though significant numbers depart coastal areas October - March. Woodlands, wetlands, parks, urban areas. One of the most abundant roadside birds around towns.

Ground Cuckoo-shrike *Coracina maxima*
[5.03 5.07]
Rare to uncommon resident and nomadic visitor. Recorded in Katherine and Southwest regions. Open woodland.

Black-faced Cuckoo-shrike *Coracina novaehollandiae*
Common and widespread. Resident, though largely absent from coastal areas October - April. Recorded in all regions. Woodlands, wetland edges, floodplains, parks, urban areas.

White-bellied Cuckoo-shrike *Coracina papuensis*
Common and widespread resident. Recorded in all regions. Woodlands, wetland edges, floodplains, riverine woodland, parks, urban areas.

Cicadabird *Coracina tenuirostris*
[1.01 1.07 1.18 3.06]
Uncommon resident, with local movement and probably some birds moving seasonally into and out of the Top End. Recorded in all regions. Monsoon forest, riverine forest, mangroves, occasionally gardens.

White-winged Triller *Lalage tricolor*
Common and widespread. At least some birds resident in southern areas, though only a visitor to coastal areas April - October, when numbers throughout the Top End increase markedly. Recorded in all regions. Open woodland, parks.

Varied Triller *Lalage leucomela*
[1.02 1.03 1.05 1.07 1.09 1.15 2.04 3.01 3.02 3.07 3.08 3.10 3.11]
Common and widespread resident, though less common away from coastal areas. Recorded in all regions. Mangroves, monsoon forest, riverine forest, parks, urban areas.

Olive-backed Oriole *Oriolus sagittatus*
Common and widespread resident, though generally absent from coastal areas October - April. Recorded in all regions. Woodlands, parks, urban areas.

Yellow Oriole *Oriolus flavocinctus*
[1.04 1.05 1.07 1.09 1.16 2.01 3.11]
Common and widespread resident. Recorded in all regions. Monsoon forest, riverine woodland, parks, urban areas, occasionally mangroves.

Figbird *Sphecotheres viridis*
[1.04 1.05 1.07 1.09 1.12 1.16 2.01 3.11]
Common and widespread resident. Recorded in all regions. Monsoon forest, parks, urban areas.

Grey Butcherbird *Cracticus torquatus*
[1.03 1.08 1.13 1.14 1.17 2.02 2.05 2.06 3.06 3.08 4.01 4.02 4.09]
Moderately common and widespread resident. Recorded in all regions. Open forests, woodlands.

Pied Butcherbird *Cracticus nigrogularis*
Common and widespread resident. Recorded in all regions. Open woodlands, wetland edges, parks. More cosmopolitan in habitat than Grey Butcherbird.

Black Butcherbird *Cracticus quoyi*
[1.02 1.03 1.05 1.07 1.09 1.15 2.04]
Moderately common coastal resident. Recorded in Darwin and Kakadu regions. Mangroves, occasionally monsoon forest, gardens.

Australian Magpie *Gymnorhina tibicen*

[4.06 4.08 5.08]

Uncommon to moderately common resident in far south of region. Recorded in Katherine and Southwest regions. Rarely seen north of Katherine. Woodlands, parks.

White-breasted Woodswallow *Artamus leucorynchus*

Moderately common resident. Recorded in all regions, though generally scarce or absent from coast areas September - April. Occupies wetter habitats than other woodswallows.

Masked Woodswallow *Artamus personatus*

[4.06 5.01 5.02 5.03 5.04 5.05]

Uncommon to moderately common nomadic visitor. Recorded in all regions, though particularly scarce near coast. Open woodland.

White-browed Woodswallow *Artamus superciliosus*

[4.06 5.01 5.02 5.03 5.04 5.05 5.07]

Uncommon to moderately common nomadic visitor. Recorded in all regions, though generally restricted to Southwestern Top End. Open woodland.

Black-faced Woodswallow *Artamus cinereus*

Common resident, widespread away from coast. Recorded in all regions, though generally absent from Darwin region. Plains, grasslands, wetland edges, and woodland adjacent to these.

Little Woodswallow *Artamus minor*

[1.12 3.06 3.10 4.06 4.09 5.01 5.03 5.04 5.06 5.08]

Moderately common visitor. Recorded in all regions, but generally uncommon near coast. Woodlands, sandstone plateaux.

Torresian Crow *Corvus orru*

Common and widespread resident. Recorded in all regions. Woodlands, settlements, refuse tips.

Apostlebird *Struthidea cinerea*

[4.08 5.01 5.02]

Moderately common resident, localised but expanding range in southern Top End. Recorded in Katherine and Southwest regions. Open and riverine woodlands, occasionally parks and gardens. Along the Victoria Highway often associated with water at cattleyards.

Great Bowerbird *Chlamydera nuchalis*

Common and widespread resident, recorded in all regions. Woodland, escarpment, mangrove and monsoon forest edges, wetlands, parks, urban areas.

Singing Bushlark *Mirafra javanica*
[1.13 1.17 2.02 2.05 5.05 5.08]
Common and widespread resident. Recorded in all regions. Grasslands, floodplains, wetland edges.

White Wagtail *Motacilla alba*
Vagrant. Records of single birds (race *leucopsis*) at Leanyer Sewage Works during the Dry Seasons of 1993 - 1996 (BARC 215).

Yellow Wagtail *Motacilla flava*
[1.10 1.13]
Uncommon to moderately common visitor, October - April, rare or absent in other months. Recorded in all regions except Southwest, though most regular in coastal and near coastal areas. Wetlands, floodplains, landward edge of mangroves, sewage ponds, rain puddles, occasionally parks.

[Citrine Wagtail *Motacilla citreola*]
A claim from the South Alligator River area in December 1981 (McKean 1982) is unconfirmed (BARC 92).

Grey Wagtail *Motacilla cinerea*
Vagrant or very rare visitor. Streams, especially in stone country, waterfalls. A lack of observers in the species' favoured habitats during the Wet Season means that birds are likely to be overlooked. Confirmed records from Kakadu, Leanyer Sewage Works March 1980 (McKean 1980b; BARC 27) and Jan 1998 (BARC 234).

Richard's Pipit *Anthus novaeseelandiae*
[1.10 1.11 1.13 1.17 2.02 4.10 5.06]
Moderately common and widespread resident. Recorded in all regions. May be some seasonal movement. Grassland, floodplains, wetland edges, occasionally parks, ovals.

Eurasian Tree Sparrow *Passer montanus*
Vagrant, presumably all records being ship assisted birds. Records from Darwin Harbour, Cullen Bay and Nightcliff.

Painted Finch *Emblema pictum*
[5.05]
Rare nomadic visitor to southern Top End, localised in suitable habitat. Recorded in Katherine and Southwest regions only. Rocky hills, escarpments.

Crimson Finch *Neochmia phaeton*
[1.11 1.12 1.13 1.16 2.01 2.06 3.07 4.03 4.06 4.08 5.02 5.03 5.05]
Common and widespread resident. Recorded in all regions. Rivers and streams, wetland edges, occasionally parks, gardens.

Star Finch *Neochmia ruficauda*
[5.01 5.02 5.03 5.04 5.05 5.06 5.07 5.08]
Uncommon and patchily distributed nomadic resident in Southwest, rare or absent elsewhere in Top End. Recorded in all regions, though many claims from Darwin and Fogg Dam regions are probably misidentified Crimson Finches. Rivers and streams, wetland edges, occasionally open woodlands, parks, gardens.

Zebra Finch *Taeniopygia guttata*
[4.06 4.07 4.08 4.10 5.04 5.05 5.06 5.07 5.08]
Moderately common nomadic resident in south, uncommon to rare or absent further north. Recorded in Katherine and Southwest regions. Reports from Darwin region are likely to be aviary escapes. Woodlands.

Double-barred Finch *Taeniopygia bichenovii*
Common and widespread resident. Recorded in all regions. Woodlands, forests, mangroves, wetlands, parks, urban areas.

Masked Finch *Poephila personata*
[1.14 2.02 2.05 2.06 3.08 3.11 4.03 4.04 4.09 5.02 5.03 5.04 5.05 5.06 5.08]
Moderately common and widespread resident. Recorded in all regions, though absent from coastal areas October - April. Open woodlands, grassy plains.

Long-tailed Finch *Poephila acuticauda*
[1.03 1.09 1.14 2.02 2.06 3.08 3.11 4.09 5.02 5.03 5.04 5.05 5.06]
Moderately common and widespread resident. Recorded in all regions. Woodlands, edges of wetlands and floodplains, parks. Apparently becoming established in some urban areas of Darwin.

Gouldian Finch *Erythrura gouldiae*
[4.01 4.02 4.03 4.04 4.06 4.08 4.09 4.10 5.02 5.04 5.05 5.06 5.07 5.08]
Rare to uncommon nomadic resident, widespread in southern half of Top End. Recorded in all regions, though extinct in Darwin region, with recent reports likely to be of aviary escapes. Open woodlands, grasslands, rivers, occasionally parks, gardens.

Yellow-rumped Mannikin *Lonchura flaviprymna*
[5.02 5.03 5.04 5.05 5.06 5.07 5.08]
Uncommon nomadic resident, mainly Southwestern Top End. Recorded in all regions. Rare in Darwin, with most records of birds in flocks of Chestnut-breasted Mannikins at feeding stations. Some claims of this species in Darwin undoubtedly refer to misidentified immature Chestnut-breasted Mannikins. Rivers and streams, open woodland, edges of wetlands and floodplains, occasionally parks and gardens.

Chestnut-breasted Mannikin *Lonchura castaneothorax*
[1.05 1.12 1.13 2.06 3.01 4.06 4.10 5.02 5.03 5.05 5.06]
Moderately common and widespread resident, recorded in all regions. Woodland, river and streamside vegetation, floodplains and wetland edges, parks, gardens.

Pictorella Mannikin *Heteromunia pectoralis*
[2.05 4.09 4.10 5.04 5.05 5.06 5.07 5.08]
Rare to uncommon nomadic resident. Recorded in all regions except Darwin, but generally rare in all but the Southwest region. Grassland, open woodland, wetland edges.

Mistletoebird *Dicaeum hirundinaceum*
Common and widespread resident. Recorded in all regions. Woodlands, monsoon forest, mangroves, parks, urban areas.

Barn Swallow *Hirundo rustica*
[1.09 1.10 2.01 2.02]
Uncommon visitor, October - April, rare or absent in other months. Recorded in all regions, but more common in northern areas. Mainly occurs around coasts, wetland areas, sewage ponds.

Welcome Swallow *Hirundo neoxena*
Rare, irregular visitor May - November. Recorded in all regions. Woodlands, plains, wetlands, sewage ponds.

Red-rumped Swallow *Hirundo daurica*
Vagrant. Leanyer Sewage Works, February 2003.

Tree Martin *Hirundo nigricans*
Common and widespread visitor, with greatest numbers March - November. Recorded in all regions. Aerial over wetlands, shores, woodlands, urban areas, sewage ponds.

Fairy Martin *Hirundo ariel*
Uncommon to moderately common visitor. Recorded in all regions, though scarce in northern or coastal areas. Wetlands, creeks, roadway drains and culverts.

Zitting Cisticola *Cisticola juncidis*
[1.11 1.12 1.13 2.01 3.01]
Moderately common resident. Recorded in all regions except Katherine. Reported from all months, though quiet and unobtrusive May - November. There may be significant movements between seasons. Grassy wetland edges, floodplains.

Golden-headed Cisticola *Cisticola exilis*

[1.11 1.13 2.01 2.03 2.06 3.07 4.08]

Common and widespread resident. Recorded in all regions. Wet or dry floodplains, wetlands, grasslands.

[Gray's Grasshopper Warbler *Locustella fasciolata* **]**

Claims of this species from Holmes Jungle in December 1979 and Harrison Dam, Humpty Doo, in January 1982 (McKean 1984), are unconfirmed (BARC 75).

Oriental Reed-Warbler *Acrocephalus orientalis*

Vagrant or very rare visitor to coastal and near coastal wetlands. One record from near Leanyer Swamp and several records, including two specimens and several captured birds, from Harrison Dam, near Fogg Dam (McKean 1983b). Wetlands with thickets.

Clamorous Reed-Warbler *Acrocephalus stentoreus*

[1.11 1.13 2.01 2.03 2.06 3.02 3.07 4.08]

Moderately common resident. Recorded in all regions. Wetlands with tall grasses and reeds, occasionally flooded forest at wetland edges.

Tawny Grassbird *Megalurus timoriensis*

[1.12 1.13 2.01 3.07]

Moderately common resident. Recorded in all regions. Wetlands, floodplains. Generally unobtrusive May - September.

Brown Songlark *Cincloramphus cruralis*

[2.02 5.05]

Rare to uncommon visitor to southern areas May - October, with occasional birds moving into northern areas. Recorded in all regions. Open woodlands, floodplains, grasslands.

Rufous Songlark *Cincloramphus mathewsi*

[2.02 2.05 3.02 3.08 4.09]

Moderately common visitor to southern Top End May - October, with a few birds moving nearer to coast. Recorded in all regions. Open woodlands, floodplains, grasslands.

Yellow White-eye *Zosterops luteus*

[1.02 1.03 1.05 1.07 1.09 1.15 2.04]

Common resident in coastal mangroves. Recorded in Darwin and Kakadu regions. Mangroves and habitats adjacent to mangroves, gardens.

Common Starling *Sturnus vulgaris*

Vagrant. Recorded only in Darwin area (McCrie 2001)

146

	Darwin	Fogg Dam	Kakadu	Katherine	Southwest	Australian Endemic
❏ Emu	●	●	●	●	●	●
❏ Orange-footed Scrubfowl	●	●	●			
❏ Stubble Quail			●			
❏ Brown Quail	●	●	●	●	●	
❏ King Quail	●	●	●	●	●	
❏ Magpie Goose	●	●	●	●	●	
❏ Plumed Whistling-Duck	●	●	●	●	●	●
❏ Wandering Whistling-Duck	●	●	●	●	●	
❏ Blue-billed Duck	●					●
❏ Freckled Duck	●		●			●
❏ Black Swan	●		●	●	●	●
❏ Radjah Shelduck	●	●	●	●	●	
❏ Pink-eared Duck	●	●	●	●	●	●
❏ Green Pygmy-goose	●	●	●	●	●	
❏ Australian Wood Duck	●					●
❏ Pacific Black Duck	●	●	●	●	●	
❏ Grey Teal	●	●	●	●	●	
❏ Chestnut Teal	●					●
❏ Garganey	●	●	●			
❏ Hardhead	●	●		●	●	
❏ Australasian Grebe	●	●	●	●	●	
❏ Little Grebe	●					
❏ Hoary-headed Grebe	●	●	●	●	●	●
❏ Great Crested Grebe	●	●	●	●	●	
❏ Tahiti Petrel	●					
❏ Streaked Shearwater	●					
❏ Wedge-tailed Shearwater	●					
❏ Wilson's Storm-Petrel	●					
❏ Red-tailed Tropicbird	●					
❏ Masked Booby	●					
❏ Brown Booby	●		●			
❏ Australian Darter	●	●	●	●	●	●
❏ Little Pied Cormorant	●	●	●	●	●	
❏ Pied Cormorant	●		●	●		
❏ Little Black Cormorant	●	●	●	●	●	
❏ Great Cormorant	●		●	●	●	
❏ Australian Pelican	●	●	●	●	●	

	Darwin	Fogg Dam	Kakadu	Katherine	Southwest	Australian Endemic
❑ Great Frigatebird	•					
❑ Lesser Frigatebird	•					
❑ Christmas Frigatebird	•					
❑ Little Bittern	•					
❑ Black Bittern	•	•	•	•	•	
❑ Nankeen Night Heron	•	•	•	•	•	
❑ White-faced Heron	•	•	•	•	•	
❑ Little Egret	•	•	•	•	•	
❑ Eastern Reef Egret	•		•			
❑ White-necked Heron	•	•	•	•	•	•
❑ Great-billed Heron	•	•	•	•	•	
❑ Pied Heron	•	•	•	•	•	
❑ Great Egret	•	•	•	•	•	
❑ Intermediate Egret	•	•	•	•	•	
❑ Cattle Egret	•	•	•	•	•	
❑ Striated Heron	•	•	•			
❑ Glossy Ibis	•	•	•	•	•	
❑ Australian White Ibis	•	•	•	•	•	•
❑ Straw-necked Ibis	•	•	•	•	•	
❑ Royal Spoonbill	•	•	•	•	•	•
❑ Yellow-billed Spoonbill	•	•	•	•	•	•
❑ Black-necked Stork	•	•	•	•	•	
❑ Osprey	•	•	•	•	•	
❑ Pacific Baza	•	•	•	•		
❑ Square-tailed Kite	•	•	•	•	•	•
❑ Black-breasted Buzzard	•	•	•	•	•	•
❑ Black-shouldered Kite	•	•	•	•	•	•
❑ Letter-winged Kite		•	•	•	•	•
❑ Black Kite	•	•	•	•	•	
❑ Whistling Kite	•	•	•	•	•	
❑ Brahminy Kite	•	•	•	•	•	
❑ White-bellied Sea-Eagle	•	•	•	•	•	
❑ Swamp Harrier	•	•	•	•	•	
❑ Spotted Harrier	•	•	•	•	•	•
❑ Grey Goshawk	•	•	•	•	•	
❑ Brown Goshawk	•	•	•	•	•	
❑ Collared Sparrowhawk	•	•	•	•	•	

	Darwin	Fogg Dam	Kakadu	Katherine	Southwest		Australian Endemic
Red Goshawk	●	●	●	●	●		●
Wedge-tailed Eagle	●	●	●	●	●		
Little Eagle	●	●	●	●	●		
Brown Falcon	●	●	●	●	●		
Nankeen Kestrel	●	●	●	●	●		
Australian Hobby	●	●	●	●	●		
Grey Falcon		●		●	●		●
Black Falcon	●	●	●	●	●		●
Peregrine Falcon	●	●	●	●	●		
Sarus Crane			●				
Brolga	●	●	●	●	●		
Buff-banded Rail	●	●	●	●	●		
Lewin's Rail	●	●					
Bush-hen	●	●	●	●			
Baillon's Crake	●	●	●				
Australian Crake	●						
Spotless Crake	●	●					
White-browed Crake	●	●	●		●		
Chestnut Rail	●		●				
Purple Swamphen	●	●	●	●	●		
Black-tailed Native-hen					●		●
Eurasian Coot	●	●	●	●	●		
Australian Bustard	●	●	●	●	●		
Red-backed Button-quail	●	●	●	●	●		
Chestnut-backed Button-quail	●	●	●	●	●		●
Red-chested Button-quail	●	●	●	●	●		●
Little Button-quail					●		●
Latham's Snipe			●				
Pin-tailed Snipe	●						
Swinhoe's Snipe	●	●	●	●	●		
Black-tailed Godwit	●	●	●				
Bar-tailed Godwit	●		●				
Little Curlew	●	●		●	●		
Whimbrel	●		●				
Eastern Curlew	●		●				
Common Redshank	●		●				
Marsh Sandpiper	●	●	●	●	●		

	Darwin	Fogg Dam	Kakadu	Katherine	Southwest	Australian Endemic
❏ Common Greenshank	●	●	●	●	●	
❏ Green Sandpiper	●					
❏ Wood Sandpiper	●	●	●	●	●	
❏ Terek Sandpiper	●		●			
❏ Common Sandpiper	●	●	●	●	●	
❏ Grey-tailed Tattler	●		●			
❏ Ruddy Turnstone	●		●			
❏ Asian Dowitcher	●					
❏ Great Knot	●		●			
❏ Red Knot	●		●			
❏ Sanderling	●		●			
❏ Little Stint	●					
❏ Red-necked Stint	●	●	●			
❏ Long-toed Stint	●					
❏ Baird's Sandpiper	●					
❏ Pectoral Sandpiper	●		●			
❏ Sharp-tailed Sandpiper	●	●	●	●	●	
❏ Curlew Sandpiper	●	●	●	●	●	
❏ Stilt Sandpiper	●					
❏ Broad-billed Sandpiper	●		●			
❏ Ruff	●					
❏ Red-necked Phalarope	●					
❏ Comb-crested Jacana	●	●	●	●	●	
❏ Bush Stone-Curlew	●	●	●	●	●	
❏ Beach Stone-Curlew	●		●			
❏ Pied Oystercatcher	●		●			
❏ Sooty Oystercatcher	●		●			●
❏ Black-winged Stilt	●	●	●	●	●	
❏ Red-necked Avocet	●	●	●	●	●	●
❏ Pacific Golden Plover	●	●	●			
❏ Grey Plover	●		●			
❏ Ringed Plover	●					
❏ Little Ringed Plover	●		●			
❏ Kentish Plover	●					
❏ Red-capped Plover	●	●	●	●	●	●
❏ Lesser Sand Plover	●		●			
❏ Greater Sand Plover	●		●			

	Darwin	Fogg Dam	Kakadu	Katherine	Southwest	Australian Endemic
Caspian Plover		●				
Oriental Plover	●	●	●	●	●	
Red-kneed Dotterel	●	●	●	●	●	
Black-fronted Dotterel	●	●	●	●	●	
Masked Lapwing	●	●	●	●	●	
Oriental Pratincole	●	●	●	●	●	
Australian Pratincole	●	●	●	●	●	
Arctic Jaeger	●					
Pomarine Jaeger	●					
Black-tailed Gull	●					
Silver Gull	●		●	●	●	
Black-headed Gull	●					
Sabine's Gull	●					
Gull-billed Tern	●	●	●			
Caspian Tern	●	●	●	●		
Crested Tern	●		●			
Lesser Crested Tern	●		●			
Roseate Tern	●		●			
Black-naped Tern	●		●			
Common Tern	●		●			
Little Tern	●		●			
Bridled Tern	●		●			
Sooty Tern	●					
Whiskered Tern	●	●	●	●	●	
White-winged Black Tern	●	●	●			
Common Noddy	●					
Black Noddy	●					
Rock Dove	●			●	●	
Emerald Dove	●	●	●	●	●	
Common Bronzewing	●		●	●	●	●
Flock Bronzewing	●	●	●	●	●	●
Crested Pigeon	●	●	●	●	●	●
Spinifex Pigeon			●	●	●	●
Partridge Pigeon	●	●	●	●		●
White-quilled Rock-Pigeon					●	●
Chestnut-quilled Rock-Pigeon			●			●
Diamond Dove	●	●	●	●	●	●

	Darwin	Fogg Dam	Kakadu	Katherine	Southwest	Australian Endemic
Peaceful Dove	●	●	●	●	●	
Bar-shouldered Dove	●	●	●	●	●	
Banded Fruit-Dove			●			●
Rose-crowned Fruit-Dove	●	●	●	●	●	
Elegant Imperial-Pigeon	●					
Torresian Imperial-Pigeon	●	●	●	●	●	
Red-tailed Black-Cockatoo	●	●	●	●	●	●
Galah	●	●	●	●	●	●
Sulphur-crested Cockatoo	●	●	●	●	●	
Little Corella	●	●	●	●	●	
Long-billed Corella	●					●
Cockatiel	●	●	●	●	●	●
Rainbow Lorikeet	●	●	●	●	●	
Varied Lorikeet	●	●	●	●	●	●
Red-winged Parrot	●	●	●	●	●	
Northern Rosella	●	●	●	●	●	●
Hooded Parrot			●	●		●
Budgerigar			●	●	●	●
Oriental Cuckoo	●	●	●	●	●	
Pallid Cuckoo	●	●	●	●	●	
Brush Cuckoo	●	●	●	●	●	
Little Bronze-Cuckoo	●	●	●	●	●	
Horsfield's Bronze-Cuckoo	●	●	●	●	●	
Black-eared Cuckoo	●	●	●	●	●	
Australian Koel	●	●	●	●	●	
Channel-billed Cuckoo	●	●	●	●	●	
Pheasant Coucal	●	●	●	●	●	
Rufous Owl	●	●	●	●		
Barking Owl	●	●	●	●	●	
Southern Boobook	●	●	●	●		
Masked Owl			●			
Barn Owl	●	●	●	●	●	
Eastern Grass Owl	●	●	●	●		
Tawny Frogmouth	●	●	●	●	●	
Large-tailed Nightjar	●	●	●			
Spotted Nightjar	●	●	●	●	●	
Australian Owlet-Nightjar	●	●	●	●	●	

	Darwin	Fogg Dam	Kakadu	Katherine	Southwest	Australian Endemic
White-throated Needletail			•			
Fork-tailed Swift	•	•	•	•	•	
House Swift			•			
Azure Kingfisher	•	•	•	•	•	
Little Kingfisher	•	•	•	•		
Blue-winged Kookaburra	•	•	•	•	•	
Forest Kingfisher	•	•	•	•	•	
Red-backed Kingfisher	•	•	•	•	•	•
Collared Kingfisher	•		•			
Sacred Kingfisher	•	•	•	•	•	
Rainbow Bee-eater	•	•	•	•	•	
Dollarbird	•	•	•	•	•	
Rainbow Pitta	•	•	•	•		•
Black-tailed Treecreeper	•	•	•	•	•	•
Red-backed Fairy-wren	•	•	•	•	•	•
Variegated Fairy-wren			•	•	•	•
Purple-crowned Fairy-wren					•	•
White-throated Grasswren			•	•		•
Red-browed Pardalote			•	•	•	•
Striated Pardalote	•	•	•	•	•	•
Weebill	•	•	•	•	•	•
Green-backed Gerygone	•	•	•	•	•	
White-throated Gerygone	•	•	•	•	•	
Large-billed Gerygone	•	•	•	•	•	
Mangrove Gerygone	•		•			•
Dusky Honeyeater	•	•	•	•	•	
Red-headed Honeyeater	•	•	•	•	•	
Banded Honeyeater	•	•	•	•	•	•
Black Honeyeater				•		•
Brown Honeyeater	•	•	•	•	•	
White-lined Honeyeater		•	•	•	•	•
Singing Honeyeater	•	•	•	•	•	•
White-gaped Honeyeater	•	•	•	•	•	•
Grey-headed Honeyeater					•	•
Yellow-tinted Honeyeater	•		•	•	•	
Grey-fronted Honeyeater				•	•	•
White-throated Honeyeater	•	•	•	•	•	

	Darwin	Fogg Dam	Kakadu	Katherine	Southwest	Australian Endemic
❑ Black-chinned Honeyeater			●	●	●	●
❑ Little Friarbird	●	●	●	●	●	
❑ Helmeted Friarbird	●	●	●	●		
❑ Silver-crowned Friarbird	●	●	●	●	●	●
❑ Bar-breasted Honeyeater	●	●	●	●	●	●
❑ Rufous-banded Honeyeater	●	●	●	●	●	
❑ Rufous-throated Honeyeater	●	●	●	●	●	●
❑ Blue-faced Honeyeater	●	●	●	●	●	
❑ Yellow-throated Miner	●	●	●	●	●	●
❑ Crimson Chat					●	●
❑ Yellow Chat	●	●	●	●	●	●
❑ Jacky Winter		●	●	●	●	
❑ Lemon-bellied Flycatcher	●	●	●	●	●	
❑ Red-capped Robin					●	●
❑ Hooded Robin			●	●	●	●
❑ Mangrove Robin	●		●			
❑ White-browed Robin		●	●	●	●	●
❑ Grey-crowned Babbler	●	●	●	●	●	
❑ Varied Sittella	●	●	●	●	●	●
❑ Crested Shrike-tit			●	●		●
❑ Brown Whistler	●	●	●	●	●	
❑ Mangrove Golden Whistler	●	●	●			
❑ Rufous Whistler	●	●	●	●	●	
❑ White-breasted Whistler	●		●			●
❑ Little Shrike-thrush	●	●	●	●	●	●
❑ Sandstone Shrike-thrush			●	●	●	●
❑ Grey Shrike-thrush	●	●	●	●	●	
❑ Willie Wagtail	●	●	●	●	●	
❑ Northern Fantail	●	●	●	●	●	
❑ Mangrove Grey Fantail	●		●			
❑ Grey Fantail	●	●	●	●	●	
❑ Rufous Fantail	●	●	●	●	●	
❑ Spangled Drongo	●	●	●	●	●	
❑ Black-faced Monarch	●					
❑ Leaden Flycatcher	●	●	●	●	●	
❑ Broad-billed Flycatcher	●	●	●	●		
❑ Restless Flycatcher	●	●	●	●	●	●

		Darwin	Fogg Dam	Kakadu	Katherine	Southwest	Australian Endemic
❏	Shining Flycatcher	●	●	●	●	●	
❏	Magpie-lark	●	●	●	●	●	
❏	Ground Cuckoo-shrike				●	●	●
❏	Black-faced Cuckoo-shrike	●	●	●	●	●	
❏	White-bellied Cuckoo-shrike	●	●	●	●	●	
❏	Cicadabird	●	●	●	●	●	
❏	White-winged Triller	●	●	●	●	●	
❏	Varied Triller	●	●	●	●	●	
❏	Olive-backed Oriole	●	●	●	●	●	
❏	Yellow Oriole	●	●	●	●	●	
❏	Figbird	●	●	●	●	●	
❏	Grey Butcherbird	●	●	●	●	●	●
❏	Pied Butcherbird	●	●	●	●	●	●
❏	Black Butcherbird	●		●			
❏	Australian Magpie				●	●	
❏	White-breasted Woodswallow	●	●	●	●	●	
❏	Masked Woodswallow	●	●	●	●	●	●
❏	White-browed Woodswallow	●	●	●	●	●	●
❏	Black-faced Woodswallow		●	●	●	●	●
❏	Little Woodswallow	●	●	●	●	●	●
❏	Torresian Crow	●	●	●	●		
❏	Apostlebird				●	●	●
❏	Great Bowerbird	●	●	●	●	●	●
❏	Singing Bushlark	●	●	●	●	●	
❏	White Wagtail	●					
❏	Yellow Wagtail	●	●	●	●		
❏	Grey Wagtail	●		●			
❏	Australasian Pipit	●	●	●	●	●	
❏	Eurasian Tree Sparrow	●					
❏	Painted Finch				●	●	●
❏	Crimson Finch	●	●	●	●	●	
❏	Star Finch	●	●	●	●	●	●
❏	Zebra Finch				●	●	
❏	Double-barred Finch	●	●	●	●	●	●
❏	Masked Finch	●	●	●	●	●	●
❏	Long-tailed Finch	●	●	●	●	●	●
❏	Gouldian Finch			●	●	●	●

	Darwin	Fogg Dam	Kakadu	Katherine	Southwest		Australian Endemic
Yellow-rumped Mannikin	●	●	●	●	●		●
Chestnut-breasted Mannikin	●	●	●	●	●		
Pictorella Mannikin		●	●	●	●		●
European Goldfinch	●						
Mistletoebird	●	●	●	●	●		
Barn Swallow	●	●	●	●	●		
Welcome Swallow	●	●	●	●	●		
Red-rumped Swallow	●						
Tree Martin	●	●	●	●	●		
Fairy Martin	●	●	●	●	●		
Zitting Cisticola	●	●	●		●		
Golden-headed Cisticola	●	●	●	●	●		
Oriental Reed-Warbler	●	●					
Clamorous Reed-Warbler	●	●	●		●		
Tawny Grassbird	●	●	●	●	●		
Brown Songlark	●	●	●	●	●		●
Rufous Songlark	●	●	●	●	●		●
Yellow White-eye	●		●				●
Metallic Starling	●						
Common Starling	●						

Literature cited

Bywater, J & McKean, J. L (1987), "A Record of Latham's Snipe *Gallinago hardwickii* in the Northern Territory", *Aust Bird Watcher* 12:65

Christides, L & Boles, W (1994), *The Taxonomy and Species of Birds of Australia and its Territories*, RAOU, Melbourne.

Deignan, H. G (1964), "Birds of the Arnhem Land Expedition", in R.L. Specht, *On Records of the American Australian Scientific Expedition to Arnhem Land*, Vol. 4: 345-425. Melbourne University Press

Higgins, P.J & Davies, S.J.J.F (1996), *Handbook of Australian, New Zealand and Antarctic Birds*, Vol.3, OUP, Melbourne.

McCrie, N (1995), "First Record of the Kentish Plover *Charadrius alexindrinus* in Australia", *Aust Bird Watcher* 16: 91-95.

McCrie, N (2000), "A Record of the Common Starling *Sturnus vulgaris* in Darwin, Northern Territory", *N.T. Naturalist* 16: 26-27.

McCrie, N (2000), "A Sighting of a Green Sandpiper *Tringa ochropus* at Darwin, Northern Territory", *Aust Bird Watcher* 18: 229-232.

McCrie, N & Jaensch, R (1999), "Sighting of a Caspian Plover *Charadrius asiaticus* at Lake Finniss, Northern Territory", *Aust Bird Watcher* 18: 81-86.

McCrie, N & McCrie, T (1999), "Sight Records of a Black-headed Gull *Larus ridibundus* in the Northern Territory", *Aust Bird Watcher* 18: 87-92.

McKean, J. L & Dampney, A. R (1984), "First sighting of the Spotted Redshank *Tringa erythropus* in Australia", *N.T. Naturalist* 7: 8-9.

McKean, J. L & Hertog, A. L (1981), "Some Further Records of Uncommon Migrant Waders Near Darwin, N.T.", *N.T. Naturalist* 4: 10-13.

McKean, J. L & Thompson, H. A. F (1983), "Northern Territory record of the Japanese Gull *Larus crassirostris*", *Aust Bird Watcher* 10: 84-85.

McKean, J. L (1980a), "A sight record of the Green Sandpiper *Tringa ochropus* from the Northern Territory", *Aust Bird Watcher* 8: 165-166.

McKean, J. L (1980b), "A sight record of the Grey Wagtail *Motacilla cinerea* in the Northern Territory", *Aust Bird Watcher* 8: 237

McKean, J. L (1982). "A Citrine Wagtail (*Motacilla citreola*) sighting from Arnhem Land, N.T.", *N.T. Naturalist* 5: 21.

McKean, J. L (1983a), "Lewin's Rail *Rallus pectoralis*, a new addition to the avifauna of the Northern Territory", *Aust Bird Watcher* 10: 137.

McKean, J. L (1983b), "Some notes on the occurrence of the Great Reed Warbler *Acrocephalus arundinaceus* in the Northern Territory", *N.T. Naturalist* 6: 3-8.

McKean, J. L (1984a), "A Northern Territory sighting of the Baird's Sandpiper", *Aust Bird Watcher* 10: 169.

McKean, J. L (1984b), "The occurrence in Australia of Gray's Grasshopper Warbler *Locustella fasciolata*", *Aust Bird Watcher* 10: 171-172.

McKean, J.L (1985), "Birds of the Keep River National Park (Northern Territory), including the Night Parrot *Geopsittacus occidentalis*", *Aust Bird Watcher*, 11: 114-130

McKean, J. L, Bartlett, M. C & Perrins, C. M (1975), "New records from the Northern Territory", *Aust Bird Watcher* 6: 45-46

McKean, J. L , Thompson, H. A & Estbergs, J. A (1976), "Records of uncommon migrant waders near Darwin", *Aust Bird Watcher* 6: 143-148.

McKean, J. L, Hertog, A. L & Marr, N (1982), "An Australian record of the Stilt Sandpiper (*Micropalama himantopus*)", *N.T. Naturalist* 5: 22-23.

Noske, R & van Gessel, F (1987), "First Record of the Blue-billed Duck for the Northern Territory", *N.T. Naturalist* 10: 13

Noske, R & Brennan, G (2002), *The Birds of Groote Eylandt*, NTU Press.

Patterson, R (1991), "RAOU Records Appraisal Committee Opinions and Case Summaries 1988-1991", RAOU Report No. 80.

Patterson, R (1996), "RAOU Records Appraisal Committee Opinions and Case Summaries 1992-1995", RAOU Report No. 101.

Robertson, D.G (1980), "First Record of the House Swift *Apus affinus* (Apodidae) in Australia", *Aust Bird Watcher* 8: 239-242.

Shannon, G & McKean, J. L (1983), "First record of Sabine's Gull *Xema sabini* from Australia", *Aust Bird Watcher* 10: 82-83.

Storr, G. M (1977). *Birds of the Northern Territory*. Perth, Western Australian Museum. (Special Publication number 7)

Wright, A, Jaensch, R, Woinarski, J & Soulos, P (1995), "First Sighting of the Elegant Imperial-Pigeon *Ducula concinna* in Australia", *Aust Bird Watcher* 16: 110 - 114

Further information & Useful Contacts

Accommodation & Vehicle Hire
The Northern Territory Tourist Commission offers information and advice on all aspects of travelling in the NT, including accommodation and vehicle rental.
Telephone: (08) 8951 8471 or 13 6110 (Freecall from within Australia)
E-mail: email.nttc@nt.gov.au
www.ntholidays.com

Road conditions
Up to date advice on all Northern Territory road conditions.
Telephone: 1800 246 199 (Freecall from within Australia)
The NT government's Department of Infrastructure, Planning and Environment website, contains information on road conditions, tidal predictions and public transport as well as other useful information for the visiting birdwatcher.
www.nt.gov.au/ipe/dtw

Leanyer Sewage Works
For access to Leanyer Sewage Works contact NT Power and Water.
Telephone: 1800 245 091 (Freecall from within Australia)

Parks & Wildlife Commission of the Northern Territory
Information on Northern Territory National Parks and Reserves.
Goyder Centre 25 Chung Wah Terrace Palmerston NT 0830
Telephone: (08) 8999 5511
www.nt.gov.au/ipe/pwcnt

NT Field Naturalists Club (NTFNC)
There is no birding club in the Northern Territory, however the Darwin based NTFNC conducts monthly meetings and outings with a general Natural History focus.
P.O. Box 39565 Winnellie NT 0830

Mary River Park (see birdwatching site 2.06, p.56)
Eco-tourism resort, along the Arnhem Highway between Darwin and Kakadu National Park. Cabin accommodation and camping facilities, bar & meals. Guided birdwatching including river cruises are available.
Telephone: (O8) 8978 8877
E-mail: general@maryriverpark.com.au
www.maryriverpark.com.au

Birdwatching Information
Niven McCrie's website contains information on the birds of the region covered in this book, including recent sightings, checklists, identification tips and photos. Bird sightings of special interest can be reported to Niven by email.
E-mail: nivenmccrie@bigpond.com
www.users.bigpond.com/birdsnt